TIME IN ANTIQUITY

Time in Antiquity explores the different perceptions of time from Classical antiquity, principally through the technology designed to measure, mark or tell time. The material discussed ranges from the sixth century BC in archaic Greece to the third century AD in the Roman Empire, and offers fascinating insights into ordinary people's perceptions of time and time-keeping instruments.

Cosmic time is defined, as expressed through the movements of the sun, moon and stars in themselves or against the backdrop of the natural landscape. Robert Hannah subsequently discusses calendars, artificial schedules designed to mark time through the year, with particular attention being given to an analysis of the Antikythera Mechanism – the most complex, geared, astronomical instrument surviving from antiquity, and the object of exciting recent scientific studies.

At the core of the book is an analysis of the development of sundial technology, from elementary human shadow-casting to the well-known spherical, conical and plane sundials of antiquity. The science behind these sundials, as well as other means of measuring time, such as water clocks, is explained in simple and clear terms. The use of the built environment as a means of marking time is also examined through a case study of the Pantheon in Rome. The impact of these various instruments on ordinary human life is highlighted throughout, as are ordinary perceptions of time in everyday life.

Robert Hannah is Professor of Classics at the University of Otago, New Zealand. His research interests include Greek and Roman archaeoastronomy, Classical art and the Classical tradition.

SCIENCES OF ANTIQUITY

Sciences of Antiquity is a series designed to cover the subject-matter of what we call science. The volumes discuss how the ancients saw, interpreted and handled the natural world, from the elements to the most complex of living things. Their discussions on these matters formed a resource for those who later worked on the same topics, including scientists. The intention of this series is to show what it was in the aims, expectations, problems and circumstances of the ancient writers that formed the nature of what they wrote. A consequent purpose is to provide historians with an understanding of the materials out of which later writers, rather than passively receiving and transmitting ancient "ideas", constructed their own world-view.

Also available from Routledge:

ANCIENT MEDICINE
Vivian Nutton

ANCIENT METEOROLOGY
Liba Taub

ANCIENT MATHEMATICS
Serafina Cuomo

COSMOLOGY IN ANTIQUITY
Rosemary Wright

ANCIENT NATURAL HISTORY
Roger French

ANCIENT ASTROLOGY
Tamsyn Barton

TIME IN ANTIQUITY

Robert Hannah

Routledge
Taylor & Francis Group
LONDON AND NEW YORK

First published 2009
by Routledge
2 Park Square, Milton Park, Abingdon, Oxon, OX14 4RN

Simultaneously published in the USA and Canada
by Routledge
605 Third Avenue, New York, NY 10017

Routledge is an imprint of the Taylor & Francis Group, an informa business

Typeset in Garamond 3 by
Swales & Willis Ltd, Exeter, Devon

British Library Cataloguing in Publication Data
A catalogue record for this book is available
from the British Library

Library of Congress Cataloging in Publication Data
A catalog record for this book has been requested

ISBN 13: 978–0–415–33155–5 (hbk)
ISBN 13: 978–0–415–33156–2 (pbk)

FOR PAT, NGAIRE AND MARK

CONTENTS

LIST OF FIGURES

ABBREVIATIONS

CCAG	*Catalogus Codicum Astrologorum Graecorum*, Brussels: H. Lamertin, 1898–.
CIL	*Corpus Inscriptionum Latinarum*, Berlin: G. Reimer, 1862–.
IG	*Inscriptiones Graecae*, Berlin: de Gruyter, 1873–.
ILS	Dessau, H., *Inscriptiones Latinae Selectae*, Berlin: Weidmann, 2nd ed., 1954–55.
OGIS	Dittenberger, W., *Orientis Graeci Inscriptiones Selectae: Supplementum Sylloges Inscriptionum Graecarum*, Hildesheim: G. Olms, 1960.
PGM	Preisendanz, K. (ed.), *Papyri Graecae Magicae: die griechischen Zauberpapyri*, Stuttgart: B.G. Teubner, 1973.
P. Hibeh	Grenfell, B. P. and A. S. Hunt (eds), *The Hibeh Papyri*. London: Egypt Exploration Fund, Graeco-Roman Branch, 1906–55.

ACKNOWLEDGEMENTS

My first thanks must go to those who initially suggested this book to me: my series editor, Liba Taub, and the then publisher for Classics and Archaeology at Routledge, Richard Stoneman. Both saw potential in a book on time-reckoning, and encouraged me to attempt the project on the back of my efforts with Greek and Roman calendars. Liba has been a marvellous editor, knowing when and how to suggest corrections or different approaches. I hope I have not disappointed them with this offering, which has grown, like any child, into something with a mind of its own. To Richard's replacement (from my perspective), Lalle Pursglove, I owe thanks for keeping me to the timetable.

Others to whom I owe a debt of gratitude include my departmental colleagues at Otago, past and present, most notably Jon Hall, who has been a fount of knowledge about Roman social life, and from whom I have gained a great deal; Robin Hankey, for covering some of my teaching; †Agathe Thornton, née Schwarzschild, daughter and sister of two famous astronomers, but whose own interests were in cosmogony, not cosmology, and whose teaching showed me there were no fixed disciplinary boundaries; Andrew Barker and John Barsby, who got me thinking about portable sundials and a sundial in Selçuk respectively; Pat Wheatley and John Walsh, for pointers on ancient postal services, and Oropos; and Arlene Allan for various time-related references. Among my students Stefan Pedersen, Kate Anscombe and Alisa Moore deserve special mention: it has been my fate not to teach much of my research at undergraduate level, so I am doubly grateful when postgraduate students choose to work in the same field with me. Stan Lusby and Karl Hart worked as Research Assistants for me, and I am grateful for their open-minded, not to say idiosyncratic, ideas. I am grateful to my teachers in the arcane art of epigraphy, most now long gone, most 'before their time': †Chris Ehrhardt, †Anne Jeffery, †David Lewis, †Peter Fraser, and finally Michael Osborne, who generously inducted me into the mysteries of the Epigraphical Museum in Athens. This background has fed into my treatment of the Oropos equatorial sundial and the Antikythera Mechanism, two of the most crucial 'time machines' surviving from Classical antiquity. I hope their efforts are not wasted here.

Roger Beck and Denis Feeney have separately increased my sensitivity to the different ways the objects of time can be interpreted. Tony Spalinger and Chris Bennett have been an unfailing source of information and ideas regarding ancient calendars. Clive Ruggles and Efrosyni Boutsikas have opened my eyes to the wider physical landscapes in which humans have viewed celestial events. Daryn Lehoux kindly shared his thoughts with me about *parapēgmata*, and it has been a pleasure to see his own work develop in this area. Tim Parkin fed me sundial references. Tony Freeth, Mike Edmunds, Yanis Bitsakis and John Seiradakis of the Antikythera Mechanism Research Group, along with Alexander Jones, have been extraordinarily generous with their data and thoughts, as we learn more and more about this unusual instrument. Michael Wright, who probably understands more about the Mechanism's engineering than any other individual, has also been most generous with his time and ideas. John Oleson gave me the opportunity to test ideas first in a chapter for his *Handbook of Engineering and Technology in the Classical World*. John Ramsey has been generous with his knowledge of ancient comets. My brothers, John and David, have helped in areas of mathematics, astronomy and Akkadian. James Harding and John Healey kindly helped me with an Aramaic inscription on a stray sundial. Emilie Savage-Smith, Michael Rogers, David King, Zur Shalev and George Majeska all helped me unravel the problem of Ulugh Beg's illusory meridian line in Hagia Sophia.

The developers of the Voyager computer planetarium (Voyager 4.1.0, Carina Software, 865 Ackerman Drive, Danville, CA 94526, USA) deserve a bouquet for the latest version of an excellent programme – the fixes to earlier bugs came at just the right time for me to incorporate my horizon of Athens; as someone who goes back to the days of calculating astronomical positions and transferring these data to hand-drawn charts, I can only look in wonder at what computerised planetaria now offer. Staff at the Shrine of Remembrance in Melbourne kindly gave me access to its inner workings. I am also grateful to staff at the British Museum (especially Peter Higgs) for allowing me access to their sundial collection. Andrew Wallace-Hadrill and Maria Pia Malvezzi worked their magic in gaining me permits to study two meridians in Rome: Danti's in the Sala della Meridiana in the Tower of the Winds in the Vatican, and Augustus' in the Campo Marzio; to Andrew I owe a personal debt for his support elsewhere. To the staff of the 1st Ephorate, Prehistoric & Classical Antiquities, in the Archaeological Services in Athens I am grateful for access to the Tower of the Winds there. I thank Hermann Kienast for sharing his incomparable knowledge of this monument with me.

Others I must thank (and I hope that friends and colleagues whose names I inadvertently omit here will forgive my lapse of memory) are: Zosia Archibald, Mary Blomberg, Alan Bowen, Bridget Buxton, Sean Byrne, James Davidson, Barry Empson, Matthew Fox, Geoffrey Greatrex, Dick Green, Goran Henriksson, Peter Heslin, David Hutchinson, Donald Kerr, †Douglas Kidd, Clemency Montelle, John D. Morgan II, Lambert Rosenbusch, Keith

Rutter, Guy Smoot, John Stenhouse, Tom Stevenson, Richard Talbert, Harold Tarrant, Martin West, Jim Williams and Mark Wilson Jones. I hope this book repays the enjoyment I have had in discussing its issues with these people. Of course, any errors that persist are my own responsibility. Unless otherwise stated, all translations are my own.

For funding I am very grateful to the Royal Society of New Zealand Marsden Fund, which gave me a major three-year grant that allowed me the wherewithal to get started on this project in earnest. To Janet Rountree, my former PhD student and then colleague, I owe thanks for taking over some of my teaching so that I could concentrate on this research; it is a delight to see her now on the staff at Otago (albeit in Computer Science!). I am also indebted to the University of Otago for a semester of Research and Study Leave in 2007, during which half of this book was written, and for funding to various conferences, at which I have first flown some of the ideas presented in this book. To the University of Otago's library staff thanks are due for their prompt, indefatigable and always friendly service in pursuit of my frequent and often obscure interloan requests (and my thanks to the various libraries who kindly sent me material). My gratitude also goes to my department at Otago for Summer School funding, which enabled me to acquire books and interloans. A special thank-you is due to Donald Kerr for access to the extraordinary Rare Books collection that Otago is fortunate to hold; and to Simon Hart for reference to web-based material on Time. The Library of the Warburg Institute in London has continued to be a treasure-trove for many years.

As ever, my greatest debt is to my wife, Pat, and children, Ngaire and Mark. We have travelled more or less together to many places – notably in Italy, Greece and Turkey – and always in midwinter (the price for being Classical archaeologists in the Antipodes), in search of temples and museums. We have usually found our way well off the beaten track, been caught in snow blizzards, clung to precipitous mountain tracks, and negotiated flooded roads. Pat has been there all along. This book could not have been written without her.

Figure 0.1 The sundial above the Theatre of Dionysos in Athens.
Source: Photograph R. Hannah

1

TIME IN ANTIQUITY

An introduction

Hell, some wit in antiquity once suggested, was made by God for those who asked what he was doing before he made Heaven and Earth. The quip is retailed by Augustine of Hippo (*Confessions* 11.12.14), in an uncharacteristically light moment in a serious disquisition about time. The point of the joke, he tells us, is that before the Creation, time did not exist, so there is no point in asking about any 'before'. Modern cosmologists face a similar problem when dealing with questions about what happened before the 'Big Bang', which currently holds sway as the best theory for the beginning of the universe. Before this event there could be no time, nor space, so the question, 'What happened before?', is just as meaningless now as it was in Augustine's day.[1]

Yet time fascinates us. According to the 11th revised edition of the *Concise Oxford English Dictionary*, the word 'time' is the most common noun in the English language. How it stands in other languages I do not know, but it may not be too different. And it does not stop there: other time-related words are high in the popularity stakes in English, with 'year' ranked third, 'day' fifth, and 'week' seventeenth.[2]

Occasionally such popular fascination with the parts of time bubbles up from antiquity too. One poor individual, who died at the hands of robbers, seemingly along with his seven foster-children, was buried by his widow with this epitaph:

> To Iulius Timotheus, who lived more or less 28 years, a man of most innocent life, deceived by robbers with his seven foster-children, Otacilia Narsica [dedicated this] to her dearest husband.
>
> (*CIL* 6.20307; *ILS* 8505[3])

The phrase 'more or less' is expressed simply as *P M* in the inscription, that is, *plus minus* in Latin. 'Plus or minus' has become part of our everyday language to express an approximation. What the epitaph demonstrates is a desire for precision – otherwise, why bother saying it? – but an inadequate means of measuring it.

A lawyer in Dalmatia, on the other hand, had his age noted on his tombstone down to the very hour: '47 years, 9 months, 7 days, in the fifth hour of the night' (*CIL* 3.2127A; *ILS* 7774).[4] This, however, is unusual. As we shall see in the course of this book, the measurement of time underwent a slow evolution, whose stages still remain unclear, and whose results are usually not very precise by our modern, artificial standards.

We live in a curious age. Despite our knowing the mechanics of the cosmos so well now in contrast to past ages, we persist in saying that the sun 'rises' and 'sets', even though the sun does no such thing. Yet we also define a 'second' now as 'the duration of 9,192,631,770 periods of the radiation corresponding to the transition between the two hyperfine levels of the ground state of the caesium 133 atom'.[5] There was a time, not long ago, when a 'second' was simply (we thought!) one 86,400th of a day, because there were 86,400 seconds in a day (24 hours = 1,440 minutes = 86,400 seconds), and that day was governed by the sun.

The apparent movement of the sun around the Earth once defined time for us. The ordinary civil day comprises the interval between one noon and the next, between successive moments when the sun is at its highest in the sky. But it became clear to scientists that this apparent movement of the sun – or more accurately, the spin of the Earth, which produces the illusion of the sun's movement – is not uniform, but is both slowing down and erratic. The frictional tidal effects of the moon on the Earth's oceans cause it to slow down, and it is erratic because of the displacement of the North and South Poles by a few metres from one year to the next. There are seasonal fluctuations also, which are due to the varying distribution of air and water across the surface of the Earth, and which cause the Earth to slow down in spring and to speed up in winter.

Modern scientific theory and practice, however, demand uniform time to very minute levels of precision. So the measurement of the civil day, and hence of its components, is inaccurate for science. Averaging out the days to produce a 'mean solar day' allows us to smooth out some of the wrinkles inherent in measuring time by the rotation of the Earth, but not to the degree of precision now required. In an effort to provide greater standardisation, from 1956 the 'mean solar second' was anchored artificially to the value it had had in 1900. This continued to prove unsatisfactory, and so from 1967 it was agreed that we should cut the conceptual umbilical cord to the rotation of the Earth, and measure time according to another system entirely – the natural vibrations of the caesium atom, which occur in the invisible, microwave part of the radio spectrum. Atomic clocks, based on this premise, have no face nor hands. Even those ancient 'clocks', the Greek and Roman sundials, which measured time through the hours of sunlight, were characterised as having a relationship to the human face: a witticism from the Roman imperial period had it that 'If you put your nose facing the sun and open your mouth wide, you'll show all the passers-by the time of day'

(*Palatine Anthology* 11.418). We have lost the human 'face' of time, but retain its language.[6]

In chapter 2 we shall return to the natural world, seeking the means by which the heavenly bodies – the sun, moon and stars – were used to help mark time for ancient Greek and Roman societies. The cycles of these luminaries can provide in themselves sufficient regularity to help people develop time-schedules or almanacs, and ultimately calendars. But I want also to put ourselves back into the physical place of Greeks and Romans and to view these same celestial bodies against the natural landscape of hills and plains, which provided a backdrop for marking special times of the year.

I have devoted the third chapter to a study of the time-schedules, calendars and cycles of antiquity. These mark moments of time, rather than measure its passage. I have structured my study around the most complex geared instrument from antiquity, the Antikythera Mechanism. This incorporates a number of the calendars and cycles of antiquity into a single ingenious, wind-up mechanism, which could predict the positions of the sun and moon, and certainly two and probably all five of the planets known to Classical antiquity. The Mechanism has been known for over a century, but it has taken some of the most sophisticated technology of modern times, such as the CT scanner, to reveal to us just how complex it is. We are familiar with this form of high-end X-ray scanning in medicine, but the research group examining the instrument have taken advantage of the facts that the Antikythera Mechanism is not a living creature which can be harmed by too much radiation, and that it cannot move of its own accord but sits perfectly still, and so they have bombarded it with far more X-rays than any organism could withstand and have at last seen through the centuries of marine decay and encrustation that it suffered in the sea off the coast of Antikythera. I must express here my gratitude to Tony Freeth, the leader of the Antikythera Mechanism Research Group, for allowing me access to the group's material and their findings before formal publication. We still do not have all the answers to the puzzles that it presents, and we may never have, but we are a considerable distance along the road towards knowing, which is all science can sometimes ever seek. To Michael Wright, who has developed over many years a full-scale working model of the same Mechanism, I owe another debt of gratitude for making the instrument come alive, and for showing something of the spirit of engineering enquiry that must have characterised the original maker. I hope he can incorporate some of the latest findings and give us further cause to wonder at the capacity of the ancient mind to magic the cosmos into a shoebox!

In the following chapters (4–6), while demonstrating the forms of various 'time machines' devised in antiquity, notably sundials and water clocks, I want also to emphasise along the way the human facet of timekeeping and time-measurement, by burrowing into the literature of the period to see what it says, here and there, about time and its effects. I want also to see what we can derive from the instruments themselves about contemporary perceptions and conceptions of time. Few people wrote about time *per se* outside the schools of philosophy, and it is not the philosophers' thoughts that I am wishing to place in centre-stage in this book, but rather the perceptions of the 'ordinary' people, who lived and worked with these instruments devised by others. We shall see also how these people reacted to these increasingly common, and even dominating, instruments, whose growing complexity demonstrates the emerging technological sophistication of the Classical world. Yet we shall see also how simple and very human some methods of telling the time were, such as the use of one's own shadow, and how human, in another sense, the very makers of these inherently complex instruments sometimes were, in not having much of a clue about the underlying theory and getting it wrong, so that they produced, for instance, sundials which could tell the hours of the day very well but which failed miserably when it came to pointing out the time of year. Designing the dials was one thing, it seems; giving the template to an artisan to manufacture (literally, by hand) a finished product and still expect accuracy, was sometimes another. That this accuracy, or its lack, was not simply a function of size we shall see in the few examples of miniature, portable sundials, which could tell the time with remarkable precision.

The final chapter provides a case study of the Pantheon in Rome, in which I float an idea about how this famous building from antiquity could be used to tell the time, and why it might do so. Here I am consciously book-ending chapter 2's emphasis on the natural landscape with the built environment, showing how both could be used as aids to mark special times.

What I do not investigate in any depth in this book is the philosophy of time as seen by thinkers from Plato and Aristotle to Proklos and Augustine. They receive occasional attention, but as it is the instruments of time that are my natural focus, and as this book sits in a series on the Sciences of Antiquity, I have chosen to focus on investigating the practical science underlying the design of these instruments, and the social impact that they had. The sociology of ancient time has taken a while to develop. This book makes an effort to increase our awareness of it.[7]

2

COSMIC TIME

Those who live in sight of clear horizons to the east or west and with few sources of artificial light are likely still to have something of the same sense of time derived from observing the heavens as the ancient world took for granted. Sunrise and sunset mark the major part of the working day, as well as signalling geographical direction to east and west. Noon is generally when the sun rides highest in the sky (barring the summer months, when many countries shift to some form of 'daylight saving' and turn their clocks forward an hour). We still talk of 'sunrise' and 'sunset', even though we know that these are illusions caused by the rotation of the earth beneath our feet and are not the result of the movement of the sun. In that sense, we are still children of the ancient world.

In the semi-rural situation in which I live in New Zealand, I can see clearly the part of the western horizon where the sun will set throughout the year (Figure 2.1). At one extreme, in midwinter, when days are shortest and nights longest, sunset occurs over a distant hill to the northwest. At the other extreme, in midsummer, when days are longest and nights shortest, the sun sets in the southwest over a fairly flat horizon.

The sun appears to move along the 70° arc between these two points on the horizon every six months, never venturing beyond it, but seeming to stand still for a few days at the midwinter and midsummer points before turning back on itself along the horizon. The situation in the east at dawn is the same, with an arc of 70° also marking out the sun's annual course between northeast and southeast along that horizon.

For those in the northern hemisphere, of course, the northern and southern limits of the sun's course represent the opposite seasons from those that I have described here for New Zealand: in midsummer the sun reaches its northernmost extreme along the horizon, while in midwinter, it reaches its southernmost points.

We can illustrate this with a view of the midsummer/midwinter arc in Athens, Greece (Figure 2.2). This time, let us take the sunrise phenomena. I have deliberately chosen a standpoint on the Pnyx, the ancient political assembly area of Athens to the west of the Akropolis – the reason will

Figure 2.1 Limits of sunset between midsummer and midwinter Dunedin, New Zealand (latitude 45°52′S).

Source: Image derived from GoogleEarth, August 2007.

become clear soon. Looking to the east, we would find the midsummer sunrise appearing in the vicinity of the peak of Mount Lykabettos, a prominent conical hill in the northeast. Six months later in midwinter, sunrise occurs over the long, high brow of Mount Hymettos in the southeast. The arc this time is about 60°.

These midsummer and midwinter points are called the solstices, from the Latin *sol* ('sun') and *sistere* ('to stand'), because the sun seems to stand still for a few days before shifting its rising and setting points back along the horizon. This term essentially describes the situation as we see it from earth, rather than what actually happens in the solar system. This geocentric perspective matches antiquity's usual view of the cosmos, and it persists in our vocabulary, despite our knowing that it is the earth that moves around the sun and not *vice versa*, simply because it captures perfectly what our senses tell us is happening.

Through the course of the seasons, then, we see the sun apparently shifting north or south along the eastern or western horizons. In the northern hemisphere, we see the sun in midsummer rising and setting at its most northerly points on the horizon. As the season shifts to autumn and winter, the sun's rising or setting point on the horizon shifts also, moving further and further south, until in midwinter it reaches its most southerly point. Then the sun

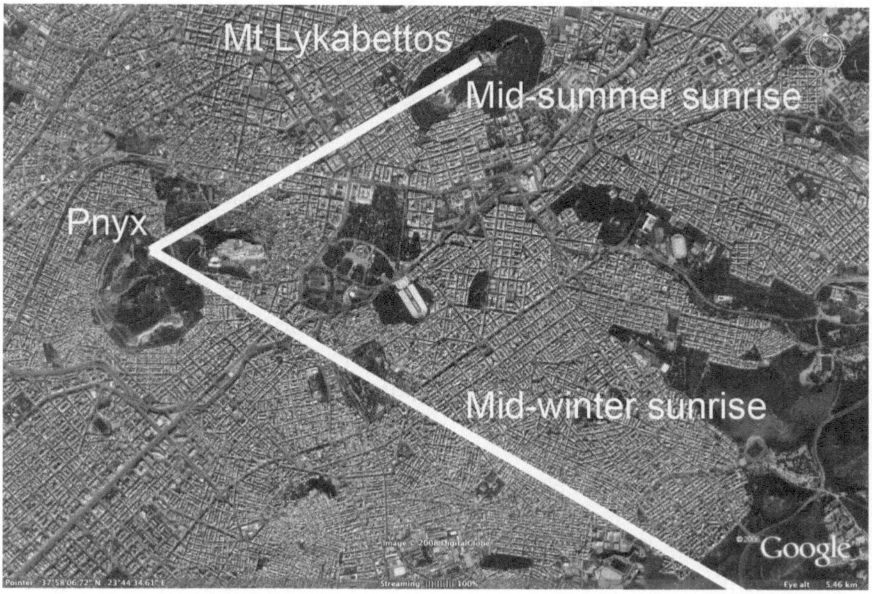

Figure 2.2 Limits of sunrise between midsummer and midwinter Athens, Greece (latitude 37°58′N).

Source: Image derived from GoogleEarth, June 2008.

returns back along the track that it has measured out on the horizon, back to the summer point. The full course, one way, takes six months; together with the return journey we have a solar year.

The mid-point between the two extremes occurs three months after midsummer or midwinter, and therefore in spring and autumn. At these points the sun rises directly in the east and sets directly in the west, and day and night are (more or less) equal. From this latter characteristic these midpoints of the sun's course are called the equinoxes, the words deriving from the Latin *aequinox* ('equal night'). Unlike the apparent standstill at the solstice periods, the equinoxes witness a rapid shift by the sun from one day to the next, so that the equinoctial points are not easy to situate with precision in either space or time.

I have tried to describe these annual phenomena in such a way as to bring out their topographical and temporal significance: the places and time at the extremes when day or night is longest, and the place and times in between when they are equal. We may surmise that other points in place and time exist elsewhere between the two extremes when the balance between day and night is different, that – like the mid-point – they occur twice each year, and that whatever balance of day and night exists on one side of the mid-point is both matched by a twin on the other side, and mirrored by its inverse at some other points. Locally for me at 46°S, midsummer day is about 15 hours 45 minutes long, while the night is 8 hours 15 minutes. Midwinter day, on the other hand, reverses these values, giving a day length of 8 hours 15 minutes, and a night-time of 15 hours 45 minutes. In between, the spring equinox provides some 12 hours 10 minutes of daylight, and 11 hours 50 minutes of night, while the autumn equinox provides about 12 hours 7 minutes of daylight, and 11 hours 53 minutes of night. In between these dates, 11 May has 9 hours of daytime, and 15 hours of night-time. This balance is matched around 1 August, on the other side of the winter solstice, while it is inverted on 5 November, as we approach the summer solstice, and on 5 February, on the other side of the summer solstice, as we leave summer.

We have to live in temperate climates, however, to notice such changes. If we live in more tropical areas, in the regions between the tropic of Cancer at latitude 23°26′N and the tropic of Capricorn at latitude 23°26′S, such as Singapore at 1°19′N or Nairobi in Kenya at 1°16′S, both practically on the equator, the difference in length between daytime and night-time throughout the year remains negligible, with almost 12 hours of day and night all the time. The Greeks and Romans were aware of such differences caused by a change in the observer's latitude (or *klima*, as they expressed it in Greek), and that this could materially affect sundial time, as we shall see in chapter 5.

When we examined the solstitial arc in Athens, I consciously chose a viewing point on the Pnyx. From that vantage-point, we found that the summer solstice sun rose in line with the local hill of Mount Lykabettos. Figure 2.3 presents a reconstruction of the view of this sunrise from the Pnyx. The Athenian horizon here provides an extremely useful topographical feature, in the form of Mount Lykabettos, to mark the point of the summer solstice. Why is this solstice particularly important? As we shall see in more detail in the next chapter, the summer solstice was the point in time from which the Athenians measured the start of their calendar year. They looked for the first new moon after this solstice and then began their first month of the year, Hekatombaiōn, from that lunar sighting. And why is the Pnyx important? For two reasons: firstly, it was the meeting place for the Athenian political assembly (the *ekklēsia*), the fundamental decision-making body of the democracy; and secondly, it was the site for a type of astronomical instrument, a *hēliotropion*, set up by Meton in the late fifth century BC. The name of the instrument in Greek suggests that it had something to do with the 'turn of the sun', which is what a solstice is, but whether it was intended to measure its placement or time, we do not know.[1]

Just how serendipitous this alignment is between the hill and an observing location in the political hub of ancient Athens for the start of the calendar year in Athens can be demonstrated anywhere. Near where I live is a similar conical hill. To achieve an effect similar to what we have met with Mount Lykabettos, I have to witness a sunset, not a sunrise, from within my home town, as the hill rises to the south (Figure 2.4).

Figure 2.3 Sunrise over Mount Lykabettos, Athens, at the summer solstice. View from the Pnyx.

Source: Chart derived from Voyager 4.1.0 (Carina Software). Horizon of Athens derived from a photograph by R. Hannah.

The sun, however, will set behind the peak of this hill not once, but twice in the year, on 27 February and 15 October, because after 27 February the sun will set progressively further to the north of the hill until the midwinter solstice in June, and then return southwards along the horizon to set behind the hill again on 15 October on its way to the midsummer solstice in December. My viewing location, a public park, is not at all significant beyond being an excellent spot to look at the local scenery, and neither of the dates, 27 February or 15 October, is significant locally, beyond representing the general limits of the university's teaching year, so the phenomenon is wasted in my city.

On the other hand, within the last hundred years or so a society has been recorded as using a phenomenon similar to the one I have posited for ancient Athens, in order to time a culturally important festival. Between 1891 and 1893, the anthropologist Alexander Stephen spent time with the Hopi Indians in the high plateaus of north-east Arizona. From a vantage point in Walpi, the Indians could look across the mesa to the southwest horizon, which was broken up by the jagged peaks of the San Francisco mountains, 135 km away. Walpi is at an altitude of about 1,750 m above sea level, while the peaks reach a height of about 3,875 m, making them still a dominant feature on the distant horizon (Figure 2.5, a view from the vicinity of the Hopi Cultural Center on Second Mesa).

Figure 2.4 Sunset behind the peak of Saddle Hill, 28 February, Dunedin, New Zealand.

Source: Photograph R. Hannah.

Figure 2.5 View of the San Francisco mountains, from the vicinity of the Hopi Cultural Center on Second Mesa, Arizona.

Source: Photograph reproduced by kind permission of Stephen McCluskey.

With the Hopi Sun Chief, Kwa'chakwa, Stephen witnessed the lead-up to the Hopi winter solstice ceremony.[2] The chief noted the sun's changing position on the peaked horizon far to the southwest, so as to know when to start the winter solstice ceremony. Stephen sketched the solar observations made by the chief (Figures 2.6 and 2.7).[3]

According to Stephen, when the sun set on 6 December along the brow of the Eldon Mesa (now Schultz Peak), to the left of the high San Francisco peaks (the three peaks, Humphreys, Agassiz and Doyle, stand out from the direction of Walpi), Kwa'chakwa predicted that the sun would set three days later 'in the notch made where Eldon Mesa intersects', that is, where Schultz Peak and the Eldon Mountain range are intersected by the Schultz Pass. After sunset on 9 December, it was agreed that the sun had reached its appropriate setting place, and that on the following morning the Crier chief, Ho'ñi, was to announce the winter solstice ceremony. Four days after the announcement, on 14 December, the Hopi started the nine-day celebrations, which culminated at the solstice on 22 December.[4]

Such use of the horizon and the sun can be replicated anywhere, but to use the coincidence so as to tell a particular time of year is, as we have seen, what is remarkable. This use of the sun and horizon does not create a sundial as such, but instead a precise suntracker for a limited part of the seasonal year. Its use in ancient Greek and Roman societies may illuminate their thinking about the sun as a timekeeper in certain contexts.

This use of the natural landscape can be paralleled by the use of the built environment. To take a non-classical and prehistoric example, at Falköping in southern Sweden over 200 Neolithic chamber tombs have been shown to

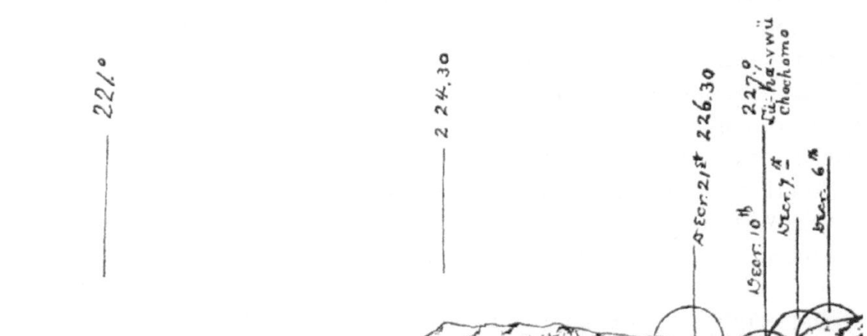

Figure 2.6 Alexander Stephen's sketch of the observations by the Hopi Sun chief, Kwa'chakwa, of the setting sun in December, over the San Francisco Mountains, Arizona.

Source: Parsons 1936: Map 4; reprinted with permission of Columbia University Press.

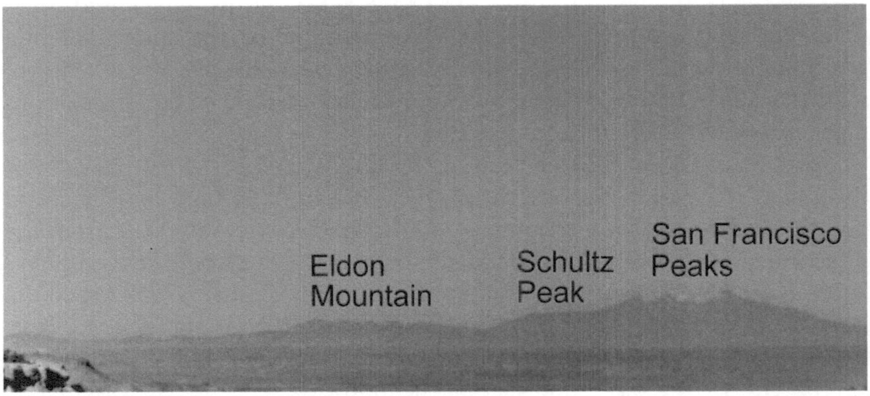

Figure 2.7 San Francisco mountains from the vicinity of Walpi, Arizona (detail of Figure 2.5).

Source: Photograph reproduced by kind permission of Stephen McCluskey.

have a striking tendency to be oriented towards sunrise on midwinter day, or towards sunrise in distinct periods of 30, 45, 60, 75 and 90 days on either side of it. Effectively this means a different orientation every 15° or so, a remarkably precise subdivision in both time and space.[5] What will have enabled this precision in orientation is the very wide solstitial arc at Falköping: at its latitude of 58°N, the solar arc between the two solstices along the horizon is 98°. This is a huge arc, capable of being subdivided by eye much more readily than its counterpart in Greece, even though the Swedish horizon at this locality is generally featureless and flat. But even in the Classical world we find noticeable orientations of public buildings, such as temples in both Greece and Rome. These have been the object of study for over a century now, but it is only recently that we have been able to discern the likeliest cause of certain orientations, through careful analysis not only of the astronomical data but also, tellingly, of the cultural data provided by relevant cult and myth.[6] We shall examine an instance of such deliberate orientations in the Roman world in the final chapter.

Our Western calendar is a solar calendar, which uses the sun as the principal means of keeping activities or events aligned with the seasons. The solar year on which this calendar depends measures the passage of time from one spring equinox to the next, and consists of 365.24219 mean solar days, or approximately 365¼ days. The odd quarter-day is absorbed into an extra single 'leap day' every four years. Ironically, the Romans, the devisers of this calendar, misunderstood this formula after its institution in 45 BC, adding the leap day by mistake every third year, until the emperor Augustus stopped the slippage by forbidding the next three leap days, and then putting the calendar on the correct footing from AD 8. The fact that the solar year is not exactly 365¼ days long, but rather 365.24219 days means that further adjustment was necessary with the reform instituted under Pope Gregory XIII in 1582, to allow for the small differences between the practical and the precise formulations of the year which accumulate over long periods of time. Further adjustment since the twentieth century, such as the insertion of the 'leap second' (as at the start of 2006), has been driven by the paradigm shift which saw us set aside the sun as the fundamental measure of time, and adopt instead the natural vibrations of the caesium atom, which in turn redefined the 'second' as the basic unit of time.

Now let us extend our observations to the stars, far smaller points of light than the sun, but more numerous and more distinctive as ornaments in the sky between the dusk and dawn. We often see them just as that, ornaments or jewels, without necessarily realising that they too move (from our perspective, anyway) through the seasons. But even if we do only glance at the night sky for a minute or two every now and then through the year, we may notice that the stars we see in one season at a given time of night in fact differ from those that dominate the sky in another season around the same time. It takes more persistent observation than we devoted to the sun's motion, and a degree of insomnia, to realise that the stars also, like the sun, circle over us in vast arcs, rising and setting through the course of the night. As with sunrise and sunset, this movement is an illusion, caused by the earth's daily rotation around its own axis. The imaginary extension of the earth's axis out into space is the axis around which the stars seem to circle. As with circles of geographical latitude around the spherical globe, these celestial circles are smallest at the northern and southern poles of the extended axis, and largest at its equator, which is simply the extension of the earth's equator out into space too.

Furthermore, from one night to the next, we can observe that any given star that rises or sets does so at gradually different times: if we see a star rise or set at, say, 6.30 p.m. one night, then we shall find that it will not rise again at exactly the same time over the following nights. This is easier to watch as evening turns to night, but it can be done also as night gives way to dawn. The stars which rise and set do so earlier each night by about four minutes, because of the earth's shift each day along the path of its orbit of the sun.

For observers closer to the earth's equator, stars closer to the North Celestial Pole will always appear to circle around the pole without ever rising from or setting below the horizon. To be more precise, stars with a celestial latitude that is equal to or greater than the co-latitude of the location of the observer on the ground will not rise or set. So if you live at 50°N, stars with a celestial latitude (called declination) between the co-latitude, 40°N (i.e. 90°–50°), and 90°N will always remain above the horizon. Stars with a declination between 40°N and 40°S will rise and set. Stars that are closer to the South Celestial Pole will not even rise above the horizon, but will always circle the pole out of sight. Again, the declinations of the stars which do this are a function of your latitude.

As we have seen, the sun rises and sets at different points within a limited arc along the horizon. In contrast, for any given location the stars rise and set over the same place on the eastern or western horizon all through the year. The bright star Arcturus in the constellation Boötes, for instance, always rises in the northeast, and sets in the northwest. So too does the cluster known as the Pleiades, or Seven Sisters, in the constellation Taurus. Alnilam, in the belt of Orion, always rises almost directly in the east and sets close to directly in the west, and thus is a good signal for the sites of the equinoctial sunrise and sunset. Antares, the prominent red star in Scorpius, rises in the southeast and sets in the southwest. The stars do this at gradually different times of the day through the year, so that for one part of the year they are visible for all or part of the night, but for another they are lost to sight in the daylight.

What this leads to is the realisation that the rising or setting of certain stars can be associated with different seasons or parts of the seasons. Stars have an advantage over the sun as timekeepers in that in any given location at a given point of the eastern horizon we can anticipate, for example, the first dawn rising of a star at a certain time of the year, or on the western horizon, say, its last evening setting. The star can thus be used as a time-marker. Where I live in New Zealand, people are becoming more aware of this utility, as the native Māori festival celebrating the first appearance of Matariki (the Pleiades, the Seven Sisters) before dawn in early June develops with celebrations and ceremonies in public contexts.[7] This first appearance before sunrise signals the traditional beginning of the Māori year.[8]

The length of a 'star year' – the period between, say, a star's first sighting before dawn and its next first sighting before dawn – is very close indeed to a solar year, so close that, within a person's lifetime, one would not notice the very slight drift that does occur between them. We noted earlier that the solar year consists of 365.24219 mean solar days, which we tend to approximate to 365¼ days for practical purposes. This year, which is technically called the tropical year, measures the passage of the sun from one spring equinox to the next. A 'star' or sidereal year, on the other hand, measures the passage of the sun across a point among the stars, and comprises 365.2564 mean solar days. Obviously this is also approximately 365¼ days, the difference between this year and the tropical year being only 20 minutes 23 seconds. Even over 100 years this difference builds up to barely a day-and-a-half. So a 'star year' is as good as a solar year for measuring the seasons. For our purposes, then, we may treat the sidereal and tropical solar years as effectively the same, so that within a person's lifetime a calendar run by observations of the stars from one year to the next is equivalent to a solar calendar.

At a certain time of the year (which is dependent upon the star's position in the sky and the observer's geographical latitude) a given star will rise at the same time as the sun and therefore be invisible because it is swamped by the sun's light. Over the next few days the star will rise earlier and earlier than the sun until it first becomes visible just before sunrise, at the end of night. For the brightest star in the night sky, Sirius, this phenomenon of first visibility occurs about an hour before sunrise; for fainter stars, it will naturally be longer before sunrise. Over the ensuing weeks the star will rise progressively earlier and earlier back through the night, until eventually it rises at the start of night, just after sunset. How close to sunset it remains visible is again a function of the brightness of the star. Then the star will disappear into the sun's light at sunset. After that, the star's rising will take place during daylight, first in the evening, then in the afternoon and finally during the morning through to sunrise, and so it will be invisible until the star reappears on the eastern horizon just before sunrise again. This progress of the star provides us with two significant phenomena: a star's first visible morning rising (often termed its heliacal rising), and its last visible evening rising (called its acronychal rising).[9] For instance, for the Pleiades, a distinctive cluster of stars which features in Greek and Roman calendars, at the latitude of Knidos (the home of a major Greek astronomer, Eudoxos) the heliacal rising currently takes place before dawn about 12 June, and its acronychal rising four months later after sunset on about 29 October.

A star's setting produces a similar sequence of phases with respect to the sun. In this case, a star will set in the west on a given day at sunrise, and therefore be invisible. Over the next few days the star will set progressively earlier than sunrise until it first becomes visible at the end of night, ahead of sunrise. Over the following weeks the star will set earlier and earlier back through the night, until eventually it sets just after sunset, at the beginning of night. The star will then disappear into the light of the setting sun, and so on through daylight – evening, midday, dawn – and so will be invisible until the star reappears on the western horizon just before dawn. This sequence provides us with two more significant phenomena: a star's first visible morning setting (called its cosmical setting), and its last visible evening setting (its heliacal setting).[10] Again if we observe the Pleiades at the latitude of Knidos at the present time, the cosmical setting takes place before dawn around 6 December, while the heliacal setting follows five months later in the evening around 4 May.

To the casual observer, the sun too seems to wheel through the day in a circle parallel to the circles tracked out by the stars at night. But to anyone observing the sun over an extended period it becomes clear that it moves not only along the horizon but also gradually across the stars, tracing its own distinctive circle, which lies aslant to the unchanging paths of the stars. We can map out this path if we look at the stars which follow the sun in the evening twilight and which precede the sun at dawn. As we have already noticed, the stars change through the seasonal year, and those which the sun seems to cross form a circle called the ecliptic. The stars along this band have long been grouped into twelve constellations, which constitute the zodiac.

This name means 'small figures' in Greek and signifies the transformation of these stars into images of 'living creatures' of animal or human form. The individual names of these zodiacal figures as they have come down to us (Aries, Taurus, Gemini, etc.) are simply Latin translations of earlier Greek names (Krios, Tauros, Didymoi, etc.), most of which in their turn are translations of the Babylonian names for these groups of stars.[11] Figure 2.8 illustrates part of the ecliptic above the horizon (the grey area at the bottom) from Virgo and Libra in the west (on the right), through Scorpius and Sagittarius in the south (in the centre), to Capricornus and Aquarius and the beginning of Pisces in the east (on the left). The solar year can be measured by the sun's apparent passage across each of these constellations, with each constellation's wider territory marking out a rough twelfth of the year. The Greeks used this method of marking out the solar year, and the Romans borrowed it from them, recognising in the twelve-part division a series of solar or zodiacal 'months'.

It is these signs of the zodiac with which we are very familiar in the west because they are still used as the basis for horoscopes in astrology. People generally are familiar with their 'birth sign', which is nowadays defined as the zodiacal 'sign' occupied by the sun at the moment of one's birth. This is a highly attenuated and fossilised form of ancient Greco-Roman astrology.[12] The fossilisation is immediately noticeable in the disengagement of the zodiacal signs from their formerly resident constellations.

If we look up our horoscopes in the media today, we look under our 'star sign', which is the zodiacal sign in which the sun was supposed to be situated at the moment of our birth. So a modern chart will tell someone born on 1 August that their sign is Leo, on the assumption that the sun was in Leo on that date. But on 1 August at present the sun is in reality situated in the constellation Cancer. Figure 2.9 illustrates the situation: on 1 August 2004 the sun in the centre of the constellation Cancer lies already above the

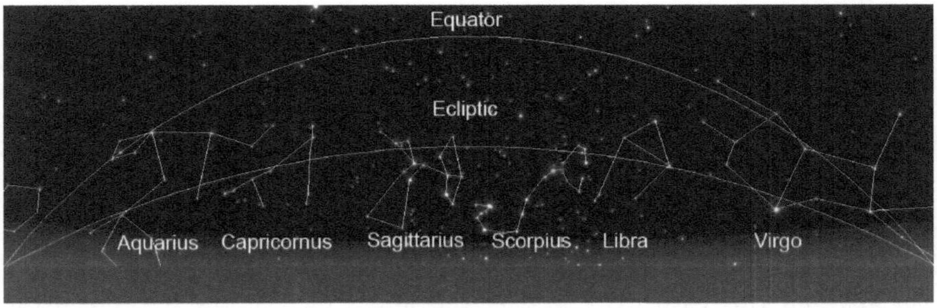

Figure 2.8 The ecliptic from Virgo and Libra in the west (on the right), through Scorpius and Sagittarius in the south (in the centre), to Capricornus and Aquarius and the beginning of Pisces in the east (on the left).

Source: Chart derived from Voyager 4.1.0 (Carina Software).

Figure 2.9 1 August 2004, soon after sunrise, with the sun in the centre of the constellation Cancer above the horizon, and Leo (with Mars, Mercury and Jupiter) just rising.

Source: Chart derived from Voyager 4.1.0 (Carina Software).

horizon, while Leo (with Mars, Mercury and Jupiter) is well behind, only just rising. The cause of the displacement lies, mechanically, in an astronomical phenomenon known as the precession of the equinoxes and, conceptually, in Western astrologers' adherence to the ancient configuration in modern times.[13]

We live in a curious age with regard to astrology. In New Zealand some time ago in a year-long, television advertising campaign, the Ministry of Transport required people to upgrade their driver's licences to a new photograph-bearing version. Technically, the upgrade had to take place within three weeks of one's birthday. But to indicate the appropriate time for the upgrade and to capture as nearly as possible the broad but constrained band of time within which it had to occur, the advertisements focussed not on clear calendar months, such as 1–31 March or 1–30 April, within which one's birthday happened to fall, but on zodiacal months. So those born 'under Aries' were told to upgrade in the 'month' of Aries (roughly 21 March–20 April), those born under Taurus to upgrade under Taurus (21 April–20 May), and so on. It appears to have been deemed more memorable for most people to identify the due date for licence upgrade with their 'star sign', and hence with a zodiacal month, than with a regular calendar month. This may say more about the New Zealand mentality than about time, but it is nevertheless a reminder of how potent these zodiacal signs remain. These advertisements highlighted an alternative means of organising time, which stems from the ancient world.

On 1 August AD 150 it was true to say, in the language of astrology, that the sun was 'in' Leo.[14] That it is no longer true in astronomical terms suggests that the stars were once seen from a different physical perspective. It is that perspective which the precession of the equinoxes has caused to change. It is best seen in the shift of stars around the North Celestial Pole. This pole is currently indicated by the star Polaris in the constellation Ursa Minor (the Little Bear). But this will not always be the case in the future, nor has it always been the case in the past. Because of the effects of the sun and moon, the earth in its spin actually wobbles very slowly like a child's spinning-top. As a result, the earth's poles themselves execute a full circle every 25,800 years. This means that what we currently observe as a Pole Star will change over a long period of time: in about 12,900 years' time, or 12,900 years ago, the North Celestial Pole was close to the star Vega in the constellation Lyra, a star over 50°, or over a quarter of the sky, away from Polaris. This effect is called lunisolar precession, or the precession of the equinoxes. As the second name implies, the stars which presently mark the position of the sun at the spring equinox have also changed over time.

The last major celestial body which affects the ancient calendar is the moon. Both it and the sun were numbered among the seven visible planets in antiquity, because they move on less regular routes than the apparently 'fixed' stars. In itself, though, the moon forms the basis of some of the principal units of time. Where the sun gives us the day, the zodiacal month and the solar or seasonal year, the moon gives us the lunar month and the lunar 'year'.

The moon completes its own orbit around the earth on average every 29.53059 days, or about once every 29½ days. It therefore shifts just over 12° across the sky every 24 hours, or about 1/15 of the dome of the sky that we see. Figures 2.10–2.12 show the movement of the moon over three successive nights, starting with it in the claws of Scorpius, then passing through the middle of the constellation, and finally leaving it to carry on to Sagittarius. If we start the moon's cycle when it is between the earth and the sun, in other words when it is 'new' and invisible, after about 14 days or so it will have moved about 180°. By then the moon lies opposite the point at which it was formerly between the sun and the earth, and now being opposite both the earth and the sun, it is fully lit up by the sun on the face it turns towards the earth, displaying itself as a 'full' moon. Midway between these two positions it has waxed to its 'first quarter' phase, and midway between the full and the next new moon, it will wane to its 'last quarter' before disappearing again. This whole period constitutes a 'month' – the word itself derives from 'moon' – and it runs usually from one new, or one full, moon to the next.

The regularity of these phases, and, generally speaking, of the months themselves to casual observers, led to the use of the month as a fundamental unit of time for ancient societies.[15] Indeed, it is initially far more important than the solar year, which is too long as a single unit of measure for practical,

Figure 2.10 Night 1: Moon in the claws of Scorpius.

Source: Chart derived from Voyager 4.1.0 (Carina Software).

everyday usage. The Classical Athenians, for instance, numbered the days of the months in three groups of ten days, which reflect the changing phases of the moon: days 1 to 10 were termed 'of the rising' (*histamenou*) as the moon waxed from new to nearly full; days 11 to 20 were simply given these

Figure 2.11 Night 2: Moon midway through Scorpius.

Source: Chart derived from Voyager 4.1.0 (Carina Software).

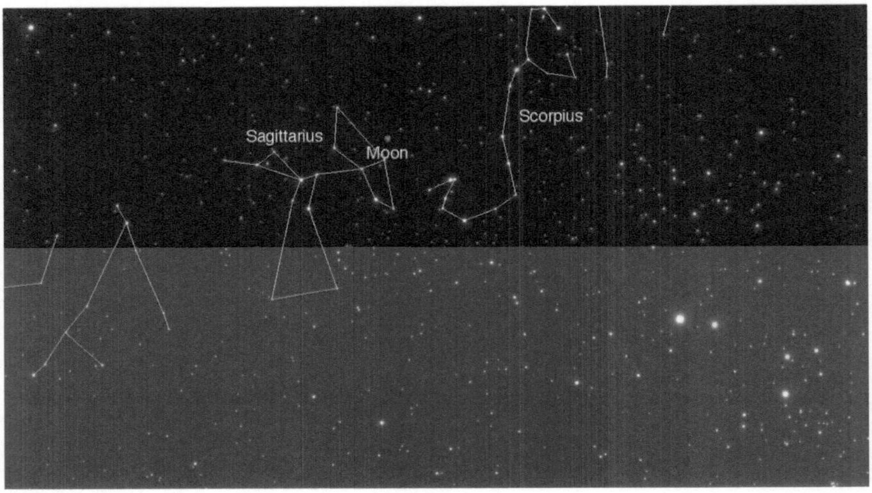

Figure 2.12 Night 3: Moon between the tail of Scorpius and the beginning of Sagittarius.

Source: Chart derived from Voyager 4.1.0 (Carina Software).

numerals; and then days 21 to 29 were numbered backwards (tenth, ninth, eighth, etc.) 'of the dying' (*phthinontos*), as the moon waned from fullish back to new. The final day of the month (regardless of whether it was a 29- or a 30-day month) was called 'the old and new' (*henē kai nea*), a term that reflects the idea that the previous evening's moon was partly the old month's moon coming to its end, and partly the new month's moon coming into being. Solon, in the early sixth century BC, was reckoned the inventor of the name (Plutarch, *Solon* 25.3; Diogenes Laertius 1.58). As Plutarch points out, the name captures the notion of the day as the one 'when one month is dying and the next is rising', an idea as old as Homer in Greek literature (*Odyssey* 14.162; 19.307).

We occasionally see this terminology for the days of the month in non-scientific contexts, which can be illuminating for popular aspects of time-keeping. Aristophanes, for instance, towards the end of the fifth century BC, has Strepsiades, a character in his comedy, *Clouds*, notice that the moon is in its 'twenties', a signal that the time has arrived when interest is due to be added to his debts (*Clouds* 16–18). The anxiety that Strepsiades feels at this realisation comes to him as he wakes up from a fitful night's sleep, and it is probably no coincidence that a moon in its 'twenties' would be a characteristic of the pre-dawn sky, rather than the evening sky. The evening sky would be the prime time to see the waxing moon.[16] The full moon is visible at both dawn and dusk, so the Athenians' three ten-day periods of the month may simply reflect the practical observation times of the prime phases of the moon: waxing in the evening sky, fullish in the evening and at dawn, and waning in the dawn sky.

The period of the full moon afforded (and still does so for us) a time when people can move about relatively freely and safely at night, as it illuminates the environment brilliantly. It is again a simple practicality of ancient life that some nocturnal activities took place only around the time of the full moon, such as parts of the Eleusinian Mysteries, held at Eleusis in mid-month.[17]

The days of the month also figure in popular perceptions of 'lucky' and 'unlucky' days. The Greek poet Hesiod in the early seventh century BC lists days of good or bad luck in the lunar months, but after him there is little in Greek literature on this issue.[18] The Romans, by way of contrast, provide a large amount of information on supposed lunar influences, particularly in relation to agricultural activity. According to this lore, crops should be planted generally just before the moon begins to wax, or during the waxing period: as the moon grows, so too, the Romans seem to have believed, would the plants. And by the same token, harvesting should take place during the waning moon.[19] According to Soranos (1.41), a Greek doctor of the second century AD, 'the ancients' believed (though he did not) that the best time for a couple to have sexual intercourse for the purpose of conceiving a child was when the moon was waxing. These 'rules' can provide some amusing side-lights on popular superstition: one should avoid having one's hair cut at the time of the waning moon, for fear of going bald (Varro, *On Farming* 1.37).

A lunar 'year' is also established, usually consisting of twelve months, or 354 days on average (12 × 29½). The trouble with such a year is that it does not sit at all well with the solar year, which governs the seasons, and which comprises about 365¼ days.

Like the solar year, a month is not a simple round number of days, but an easy way to even it out is to make successive months alternately 30 and 29 days in length, so that they average out to the requisite 29½ days. The Greeks and Romans seem not to have latched on to this method for some time. The 30-day month has a very long history among the Greeks, although it is not clear that it was always used in that time as the measure for every month. The Roman method of ironing out the creases between the lunar and solar periods is so odd that we must conclude that from an early period they divorced their idea of a 'month' from its parent, the moon.

Geminos, writing around the middle of the first century BC,[20] mentions the 30-day month at the same time as he describes the first method devised by the Greeks to make the months correlate with the solar year (*Introduction to Astronomy* 8.26). It takes more than twelve but less than thirteen months of 29½ days on average to add up to a solar year of 365¼ days, and we shall see in the next chapter some of the methods used by the Greeks to make this happen; by that stage (the early sixth century BC) alternate months of 29 and 30 days appear to be the norm. It is not clear whether the 30-day month had been discarded as the sole measure for a month in the calendar before then, but even afterwards it persisted in popular usage.

According to Geminos, the earliest means devised by the Greeks to try to make the lunar months correlate with the solar year was to insert ('intercalate') an extra month every two years. If the 30-day month was the norm, this intercalary system would give alternating years of 360 and 390 days, or 750 days for the two-year period. This biennium overruns two proper solar years by 19½ days (365¼ × 2 = 730½ days). On the other hand, if the extra month was added to two lunar 'years' of twelve months each, and if these months were alternately 30 and 29 days in length, the biennium would amount to 738 days. This overreaches two solar years by only 7½ days. This slightly better match between the lunar and solar cycles may argue for alternate 29- and 30-day months even at an early stage of Greek calendar-making. Since the historian Herodotos (2.4) in the late fifth century BC still talks of the use of the biennial system of intercalation among the Greeks of his own time, when alternating 29- and 30-day months were in use, the biennium of 738 days, rather than 750 days, would again seem more likely, and along with it the use of alternating months of 29 and 30 days.[21]

Yet Herodotos himself uses a standard 30-day month when calculating the length of a human lifespan. In a conversation that he reports (1.32) between the Athenian statesman Solon and Kroisos, the king of Lydia, Solon equates the 70 years of a lifespan initially with 25,200 days, that is, 70 years × 12 months each year × 30 days each month. Then he adds to this sum the missing intercalary months, at the rate of one month every two years. There would be 35 of these intercalary months, producing a further 1,050 days (35 × 30). These raise the total for the 70 years to 26,250 days. Throughout these calculations, all the months are 30 days long, and the ordinary, unintercalated year must be 360 days, while the intercalary year is 390 days. So either Herodotos is simply reflecting an archaic practice from Solon's time, or perhaps the 30-day month and with it the 360/390-day biennium were the norm even well into the fifth century BC in Herodotos' own time.

That it lasted even beyond then, at least for accounting purposes, is demonstrated by other evidence. In the late fifth century BC, for instance, the medical writer Hippokrates has 'four tens of seven-day periods' (i.e. 280 days) equalling nine months and ten days, implying nine 30-day months (Hippokrates, *On Flesh* 19.27–8).[22] Inscriptional evidence in the later fourth century BC shows that wages were based on the notional 30-day month: for instance, pay of 2 drachmae per day over 13 months amounts to 780 drachmae (i.e. 2 drachmae × 30 days × 13 months) (*IG* II^2. 1673, line 60; dated to perhaps 327/6 BC).

The Romans, on the other hand, had months which derived from lunar aspects, but which soon seem to have become divorced from the moon entirely. Each month was divided into three parts: the Kalends (*kalendae*) on day 1; then the Nones (*nonae*) at day 5 (in the shorter months) or 7 (in the longer months); and the Ides (*idus*) on day 13 (in the shorter months) or 15 (in the longer). Varro (*On the Latin Language* 6.27), writing around the time of the reform of the Roman calendar by Julius Caesar in 46 BC, and Macrobius (*Saturnalia* 1.15.9–11), writing in the early fifth century AD, both derive the word *kalendae* from the Greek verb *kalō* ('I call'). According to Macrobius, originally a minor priestly official was delegated the task of watching for the first sign of the new moon and then reporting its appearance to the high priest. A sacrifice would then be offered, and another priest would summon the people and announce the number of days that remained between the Kalends and the Nones, 'and in fact he would proclaim the fifth day with the word *kalō* spoken five times, and the seventh day with the word repeated seven times'. The first of the days thus 'called' was named *kalendae* after *kalō*.

All three divisions of the Roman month probably represent at least notional lunar phases: at the new moon (*kalendae*); the first quarter about a week later (the name *nonae* signifies eight days – nine by Roman inclusive reckoning); and the full moon (the name *idus* may stem from a Greek word for the full moon, as Macrobius reports among other derivations (*Saturnalia* 1.15.14–17)).[23] These characteristic divisions of the month feature on a perpetual Roman calendar, part of which has recently come to light in excavations at the Roman fort of Vindolanda in Northumberland. Only the month of September is preserved from a circular bronze disc originally about 25cm in diameter. The month is marked with K for the Kalends (1st), N for the Nones (5th), ID for the Ides (13th) along with AE for the equinox at the 23rd.[24]

According to Macrobius (*Saturnalia* 1.13), Numa, the second king of Rome (thus taking us back theoretically to the eighth to seventh centuries BC), made the city's calendar lunar by increasing the Roman year first to 354 days and then to 355, and divided the year into twelve months. The length of the year was increased to 354 days to match the time 'in which twelve circuits of the moon are completed', but then Numa afterwards added an extra day 'in honour of the odd number'. Until this time the calendar was supposed to be of ten months' length, but Numa now added January, which was made the first month of the year, and February to follow it ahead of March, where the year previously had begun. The rule of the odd number extended also to the lengths of the months. Numa organised these so that each, except February, contained an odd number of days – 29 for January, April, June, Sextilis, September, November, December, and 31 for March, May, Quintilis and October; February had 28 days.

Since the intention, we are told, was to make the year align with the moon, it is odd that the Romans are said to have opted not for a pattern of alternating months of 29 and 30 days, which copes reasonably well with the vagaries of the moon's cycle, but, because of a superstitious regard for odd numbers, for a mixture of 29- and 31-day months, excepting one month of

28 days. In reality, we may be seeing here a telescoped version of two separate stages in the development of the Roman calendar, first to a lunar one, then to a lunisolar one in which the months are more or less divorced from their lunar origins.[25]

We can tie together several of the astronomical strands covered in this chapter via a passage from Roman comedy. In his play, *Amphitruo*, produced some time around 200 BC, Plautus has the slave Sosia describe the extremely long night during which Alcumena is being impregnated by Jupiter while her husband, Amphitruo, is absent; Alcumena's child will be the superhero, Hercules:[26]

> See, the Great Bear isn't moving anywhere in the sky, and the moon doesn't change, but she's just where she was when she first rose, and Orion and the Evening Star and the Pleiades aren't setting either. Yes, the constellations are standing still, and night's not making way for day.
>
> (Plautus, *Amphitruo* 273–6)

In the context of the play, Sosia is looking for the end of night, but not finding it. He looks to the stars in the prolonged night sky and finds, in fact, that nothing has moved on. But from what point in the night have the stars stood still: from dusk, or from just before dawn, or from some time in between? The answer lies in the stars.

The planet Venus, characterised here as the Evening Star, is sandwiched by Plautus between Orion and the Pleiades, and all are failing to set. In reality the arrangement of the stars and planet – the Pleiades, Venus and Orion – and their placement near the western horizon is not impossible. The Pleiades and Orion could be observed as setting at dawn or dusk on two occasions in any year. The intrusion of Venus between them at the same time is certainly possible, because the Pleiades and Orion lie either side of the ecliptic, so Venus could – and does – traverse the space between the two. This conjunction is also a relatively common occurrence, happening two or three times every decade. It was perfectly possible for anyone on a good night, at a set time in the year and at a certain stage in the orbit of Venus, to see after twilight, an hour and a half after the sun had set, the Pleiades, Venus and then Orion all set in succession (Figure 2.13).

Whether anyone felt the need physically to scan the sky is another matter, for there were available from the fifth century BC onwards in various forms written almanacs, both popular and scientific, which reported the times of rising and setting of prominent stars at dawn and dusk throughout the year; we shall examine these in more detail in the next chapter.[27] The very stars which we are considering here – the Pleiades and Orion – figure prominently in these astronomical 'calendars'. Venus does not appear in them, nor do any of the other planets, for their positions are not fixed, but rather they alter against the background of the stars. But the absence of Venus need not

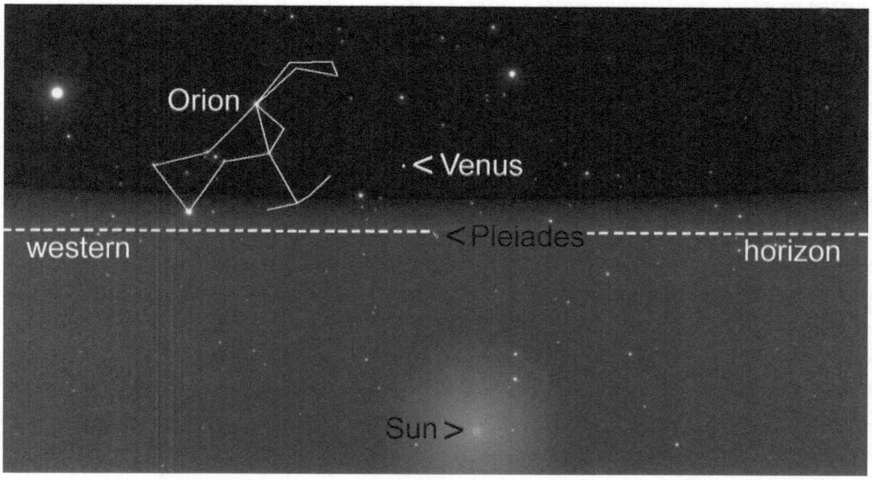

Figure 2.13 Evening setting of the Pleiades, Venus and Orion.

Source: Chart derived from Voyager 4.1.0 (Carina Software).

concern us, for her position on the ecliptic ensures that at some stage she was bound to cross the space between the Pleiades and Orion, and in fact, as I mentioned earlier, this phenomenon recurs every three or four years.

In the astronomical almanacs, we find that the settings of the Pleiades and Orion were significant events in the evenings of the months of March/April, and in the mornings of the months of October/November. So the settings of these stars existed in the Graeco-Roman consciousness outside of their mention by Plautus. They were part of the Greeks' and Romans' way of telling the time, and are not picked out at random by Plautus.

The constellation of the Great Bear, in antiquity as now for the Mediterranean and more northerly latitudes, did not rise and set at all in the night sky. It was, and is, circumpolar, perpetually wheeling around the North Celestial Pole. The movement in the Great Bear that is sought by Sosia was a part of a circle, in fact for a night in April or November almost a half circle. These characteristics warn us that Sosia must have been looking for the large and obvious motion traced by the constellation over a large part of the night, such as from dusk to dawn. For to look at the Great Bear just before dawn and to say that it is not moving is a nonsense: the movement in its pre-dawn position would be imperceptible, since unlike the Pleiades or Orion the Great Bear could not set and so move out of view altogether. This suggests that Sosia's observation reflects an evening context, not one at dawn.

The moon too, we are told, has not changed since she rose. Now, the moon can in fact rise at any time of the night, depending on its position in its orbit around the earth. It may rise at sunset, as a full moon, and then at any time between dusk and dawn as a waning moon, until it rises just before dawn about two weeks later as a slender crescent, eventually disappearing temporarily against the sun as a new moon. From then on until the next full moon, it is observed as a waxing moon rising in the daytime and setting in the western sky progressively later and later after sunset. As with the sighting by Sosia of the Great Bear, an observation that the moon's position is not changing from what it was when she first rose in the pre-dawn sky makes little or no sense. We might argue that he sees a waning crescent and so expects the sun to appear very soon. It is not her size, however, that he refers to, but her position, and this suggests that he is expecting some noticeable displacement in the moon's position. An initial sighting by Sosia of the moon as she rose at dusk and a further observation when he expected day to begin would have elicited the greatest possible change in the moon's position, from an easterly position at dusk to a westerly one at dawn. It is this large and again obvious change, I suggest, or something close to it that Sosia is asking us to imagine: instead of seeing the moon heading off to the west, he sees it still rising in the east as it had done when he first saw it. This, in turn, points to a fullish moon.

The combination of a full (or nearly full) moon and of Venus as the Evening Star that I am proposing to read in these lines of Plautus fits well with the sexual aspect of the play. The Evening Star was most commonly associated by Greek and Roman poets with love-making.[28] So too at times was the full moon, to judge from the reference by the Greek poet Pindar to the goddess Thetis' lying in love with the mortal Peleus on an evening of the mid-month, and therefore of the full moon: '. . . but on an evening in mid-month, let her loosen the lovely girdle of her maidenhood for the hero.' (Pindar, *Isthmian Ode* 8. 92–93) The child this time is the hero Akhilleus, the bane of Troy.

We may add to these observations the further point that in some Greek and Roman minds there was a belief that the best time for a couple to conceive a child was when the moon was waxing. This is mentioned (although not accepted) in the second century AD by Soranos (1.41) who ascribes the belief simply to 'some of the ancients'. All of these ideas come together when we recall the purpose of Alcumena's long night with Jupiter – the conception of the hero Hercules. The Evening Star and the moon are both present to make the occasion fertile and propitious.

So Sosia's references to the stars are intended to make the audience think of the evening sky, not of the dawn sky. Alcumena's night has been so prolonged that it seems barely to have started. In other words, the cosmic 'clock' has been stopped by Jupiter for a whole night.

3

MARKING TIME

Greek and Roman calendars variously made use of the cycles of the sun, moon and stars.[1] The moon formed the basis of all Greek city-state festival calendars, and originally underlay the main divisions of the Roman months. 'Star calendars' helped time agricultural activities in both the Greek and Roman worlds, and – in the form of the *parapēgmata* – could have assisted in regulating some Greek civil calendars. The sun, after initially loosely helping mark out seasonal periods in the agricultural cycle, eventually formed the basis of the Roman civil calendars and from them, in a nice touch of reciprocity, many Greek civil calendars.

A year measured by the sun and the stars is practically of the same length, at least over an individual's lifetime. As we saw in the previous chapter, however, a lunar 'year' – mapped out by twelve lunar months – is always incommensurate with a seasonal or solar year. A solar year consists of 365.24219 days (to express the problem in modern notation), while one lunar month averages 29.53059 days. It is therefore impossible to integrate a round number of lunar months into a single solar year: the solar year consists of more than twelve but less than thirteen lunar months. Various lunisolar cycles were devised to realign the lunar calendars with the sun, and hence with the seasons, a matter of great importance in societies whose religious lives were agriculturally based. It is, in fact, a fundamental datum of ancient Greek and Roman calendars that they derive from a desire – a need, even – to coordinate activities with nature and the gods. This means not only what may be deemed obvious and is therefore expressed explicitly – that agricultural festivals should be coordinated with their appropriate seasons in the year – but also what is less obvious and therefore not always expressed – that some activities must be avoided on certain days of the year sacred to the gods, such as political or judicial meetings.

This is a well-known aspect of the Roman calendar, for instance. Here the sacred day (*dies festus*) was distinguished from the secular (*dies profestus*), on which public and private, as opposed to the gods', business could be conducted, although some days were mixed (the *dies intercisus*). *Fasti* were lawcourt days, on which the formula for judgement ('I grant, I pronounce, I award' – *do, dico, addico*) could be pronounced, while 'non-court days' (*nefasti*) were those on which cases requiring this formula could not be held. 'Assembly days' (*comitiales*) were those days when a motion could be brought before the people (Varro, *On the Latin Language* 6.30; Macrobius, *Saturnalia* 1.16.2–5, 14–15). Figures 3.1 and 3.2 illustrate remains of two public Roman calendars from the late Republic and early Empire. Particularly prominent in the vertical columns of each month are the days marked C (*comitialis*), with N (*nefastus*) and F (*fastus*) days sprinkled among them.[2]

In this chapter it is the process of establishing time schedules that we are interested in. As our entrée into the Greek and Roman means of marking time, we shall examine a remarkable instrument which survives from the Hellenistic period, when developments in Greek astronomy started to move in a different direction from that of the Near East. All three aspects of the cosmos mentioned above – sun, moon and stars – appear in various forms on this instrument.

The Antikythera Mechanism

In 1901 a Greek sponge-diver serendipitously came across the wreck of an ancient ship at the bottom of the Mediterranean off the coast of the small island of Antikythera. The eventual recovery of the contents of the wreck, the first concerted underwater excavation, brought to the surface most notably a collection of Greek sculptures, in both bronze and marble, which are now displayed in the National Archaeological Museum in Athens. But also among the finds was a mass of bronze plates bonded together, partially

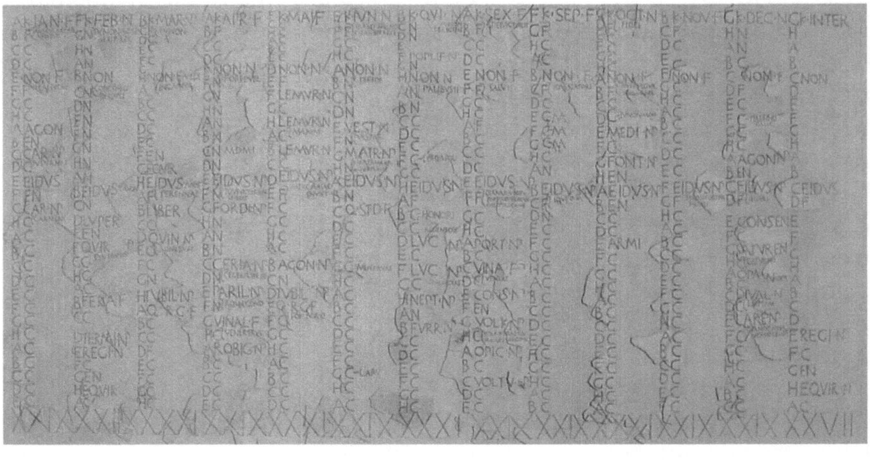

Figure 3.1 Rome, Museo Nazionale Archeologico, Palazzo Massimo: Fasti Antiates Maiores, 84–55 BC.

Source: Photograph R. Hannah.

Figure 3.2 Rome, Museo Nazionale Archeologico, Palazzo Massimo: Fasti Praenestini, AD 6–9.

Source: Photograph R. Hannah.

obliterated by corrosion or obscured by marine accretions. Cleaning and conservation showed it to be a scientific instrument of some kind, multi-geared and marked with inscriptions in Greek (Figure 3.3).

Not until X-rays were taken in the early 1970s did it become clear just how sophisticated the instrument was, with over thirty intricately interlocking gears identified and several plates interrelated by their capacity to mark time in various ways. As Derek Price demonstrated, the Antikythera Mechanism, as it has come to be called, managed to correlate the motions of the sun and the moon, timed against the twelve signs of the zodiac, the Egyptian calendar and a star calendar (a so-called *parapēgma*).[3] In the past twenty years or so this unique artifact has been studied again. Michael

Figure 3.3 Athens, National Archaeological Museum: the Antikythera Mechanism, Fragment A.

Source: Photograph R. Hannah.

Wright and Alan Bromley combined as collaborators, working primarily from the 1970s' X-rays. Wright has also worked solo, manufacturing one of the most detailed physical reconstructions of the Mechanism and explicating its underlying theory. The Antikythera Mechanism Research Group is responsible for the most remarkable discoveries to date. It comprises three teams from the UK, Greece and North America, led by Tony Freeth and Mike Edmunds: the academic team (Mike Edmunds, Tony Freeth, John Seiradakis, Xenophon Moussas, Yanis Bitsakis and Agamemnon Tselikas); the Hewlett-Packard team (Tom Malzbender, Dan Geld and Bill Ambrisco); and the Museum team (Eleni Mangou and Mary Zafeiropoulou from the National Archaeological Museum in Athens).[4] Alexander Jones and John Steele have also collaborated with the Group. Several reconstructions, both solid and virtual, have been generated as a result of these years of study.[5]

The Antikythera Mechanism Research Group subjected the Mechanism to much more powerful and subtle techniques of analysis, notably high-resolution X-ray tomography and reflectance imaging techniques, as a result producing greatly increased data about its gearing and functions.[6] We now know that the cycles of at least two, and perhaps all five, of the planets known to antiquity were tracked by the instrument's geared plates, that the motions of the sun and moon (which were also counted as planets then) were correlated using a lunisolar cycle linked to a civil calendar, and that eclipses of the sun and moon could be accurately calculated by the Mechanism.[7]

This Mechanism has usually been dated to the mid-first century BC at the latest.[8] More recent studies, however, have suggested a second-century BC context, and originally posited an association with the great contemporary Hellenistic Greek astronomer, Hipparkhos, but they have now shifted to proposing a heritage from the third-century Syracusan scientist, Arkhimedes.[9] The second-century date seems plausible, on the basis of the letter-forms of the script on the Mechanism, which bear strong similarities to inscriptions from the second half of the second century BC.[10] We should exercise caution, however, with the supposed links with Hipparkhos or Arkhimedes until firmer evidence appears.

For what purpose the Mechanism was made remains unknown. It will help, however, if we understand what its geared functions sought to achieve. The instrument's correlation of the motions of the sun, moon and planets all into one device, as well as its use of different forms of time schedules, two solar (the zodiac and the Egyptian calendar), one sidereal (the so-called *parapēgma*), all suggest that a fundamental function was not telling time, as with a clock, nor measuring time, as one does with a stopwatch, but marking or finding time, as one does with a calendar.[11]

The Metonic cycle

As I have already stated, the historical calendars of Greece and Rome display a fundamentally agricultural and religious character. They were not only created around the gods and their festivals, but (one assumes) were also meant to bind festivals and their associated rituals to the right time of the seasonal year. Yet months were initially moon-based, temporal constructs, running typically from one new moon to the next. If the agricultural festivals were to maintain alignment with the appropriate seasons, and yet also to continue to fall within the correct lunar month, some means of coordinating the lunar and solar cycles was necessary. Otherwise, these festivals would soon become divorced from their original agricultural contexts and run throughout the year every 33 years or so.[12]

Societies which run systems of reckoning time based on the moon and yet wish also to associate, say, propitiatory or thanksgiving festivals with the seasonal year face the fundamental difficulty of equating the incommensurate periods of the lunar and solar cycles. What societies discovered early on, however, is that while any given solar year cannot contain a whole number of lunar months, it is nevertheless possible to gain approximate equality between the two cycles over a period of several solar years. This is achieved by recognising that a whole number of lunar months can be assigned to a certain number of solar years, with varying degrees of accurate fit. The resultant system is a lunisolar one. In such a system, most lunar years are given twelve complete lunar months, but an occasional year will need to have a full thirteenth month added (intercalated). The trick was to discover how many lunar years, with how many lunar months, provided the least discordance with the seasons over an acceptable period of time.

Intercalary systems seem to have existed from the dawn of the historical period in Greece. Around 700 BC Hesiod refers to only one month by name in his poem, *Works and Days*, the month of Lēnaiōn, but as this is equated with the worst part of winter – 'bad days, real ox-flayers' (*Works and Days* 504) – it is possible that some form of regulatory system, however loose, had already been instituted so as to keep this lunar month within the broad season of winter. If there had not been an intercalation every now and then, the month of Lēnaiōn would have moved steadily through the seasonal year, exactly as months do in the modern religious calendar of Islam. Just as Ramadan may be associated now with summer, and now with winter, so also Lēnaiōn would have necessarily slipped through the seasons over a period of just over 33 years. A brake is needed to stop the month moving too far out of any given season. This would have to take the form of the addition of an extra month every now and then to allow the seasonal year to catch up with the shorter twelve-month lunar year, which, being shorter, is completed sooner and runs ahead.

The Greeks of Hesiod's time could have used a correlation between a star and the moon, such as we find deployed in Babylon, or between the moon and a solar solstice.[13] But even in Babylon, intercalation was an *ad hoc* event, regulated from the second millennium BC only by a limitation on which months could be doubled (the sixth and the twelfth), and by royal decree.[14] We cannot judge who would have ordered an intercalation, or suppression, of a month in Greece in Hesiod's time; nor is it clear that there was any limitation on which month could be doubled. In the historical period in Athens, for instance, we have evidence for any one of five different months being doubled as occasion demanded.[15]

From a variety of intercalary systems which were developed in Greece two deserve special mention: the *octaetēris*, and the Metonic cycle.[16] The former serves as the basis for the latter, while the latter, probably with its refinement by Kallippos, is the system used by the Antikythera Mechanism to coordinate the solar and lunar cycles.

If Censorinus (*On the Birthday* 18.5) is to be trusted in his attribution of the *octaetēris*, or eight-year cycle, to Kleostratos, it was invented by the end of the sixth century or early fifth century BC.[17] This intercalary system allowed for the regular addition of three 30-day, lunar months in three of its eight lunar years. The extra months were usually added in years 3, 5 and 8 of the cycle. Table 3.1 gives a hypothetical sequence:

Table 3.1 The *octaetēris*

Month	*Y1*	*Y2*	*Y3*	*Y4*	*Y5*	*Y6*	*Y7*	*Y8*
i	30	30	30	30	30	30	30	30
ii	29	29	29	29	29	29	29	29
iii	30	30	30	30	30	30	30	30
iv	29	29	29	29	29	29	29	29
v	30	30	30	30	30	30	30	30
vi	29	29	29	29	29	29	29	29
vii	30	30	30	30	30	30	30	30
viii	29	29	29	29	29	29	29	29
ix	30	30	30	30	30	30	30	30
x	29	29	29	29	29	29	29	29
xi	30	30	30	30	30	30	30	30
xii	29	29	29	29	29	29	29	29
Intercalary			30		30			30

In this table, columns Y1–Y8 are years in the eight-year cycle, each comprising months i–xii, alternately of 30 and 29 days, plus an intercalary month of 30 days to be set somewhere in years 3, 5 and 8 (but not usually at the end of a year). If we calculate this out, we find that eight solar years amount to 2,922 days (365¼ × 8), while eight lunar years, each of twelve months alternating with 29 and 30 days (in order to approximate the average lunar month of 29½ days), together with three extra 30-day months, also add up to 2,922 days [{(6 × 29) + (6 × 30)} × 8 + 90]. Geminos (*Introduction to Astronomy* 8.27–31) makes a similar calculation, and gives the following explanation of the *octaetēris*: each lunar year is 11¼ days behind the solar year; multiplying this difference by 8 produces 90, a round number of days, which may be divided into three whole, 30-day months, which in turn must be added to the eight years of the cycle to bring the lunar calendar back into line with the solar, from which it has rushed ahead by this amount of time.

Censorinus (*On the Birthday* 18.6) notes that many Greek cults celebrated their festivals at this interval of eight years. The Pythian Games at Delphi are mentioned expressly in this context, and the Olympic Games also can be demonstrated to have been organised according to an eight-year cycle. Both festivals, it should be noted, were actually celebrated every four years, but the ancient testimonies regarding the timing of the Games in any given cycle are strongly indicative of an overarching eight-year system.[18]

The correspondence of 2,922 days between eight solar years and the eight-year lunar cycle looks perfect. In fact it is not, because we have used approximations of both the solar year and the lunar month to arrive at the total. More precisely, our calculation for the *octaetēris* should read 2921.93752 days for eight solar years, and 2923.52841 days for the equivalent 99 lunar months. The difference is just over a day-and-a-half per *octaetēris*, with the lunar calendar running ahead of the sun by that amount every eight years.

This difference of a day-and-a-half would, of course, increase over time if no adjustments were made. After nine *octaetērides* – 72 years, or a very good lifetime in antiquity – the difference amounts to a bit more than 14 days. If the lunar calendar indicated a date 14 days ahead of the sun, what does that mean in the context of everyday life?[19] Probably not much: from what we know of the methods of Babylonian intercalation, it would take a whole month's difference between lunar and, in that case, stellar timetables to trigger action.[20] Stellar and solar, and hence seasonal, timetables are practically equivalent within a person's lifetime. There is no reason to think the Greeks were different from their eastern neighbours.

A month's difference would be accumulated after another nine *octaetērides*, or 144 years, by which time the lunar calendar would be running over 28 days ahead of the sun. At this stage, we might anticipate an extra intercalation into the *octaetēris*, but only if several generations of knowledgeable people had been keeping an eye on the glacially slow build-up of the discrepancy between the lunar calendar and the seasons. At this rate of slippage, the lunar calendar would be a whole season out of kilter only after more than 400 years. But such long stretches of time, comprising several generations, were of little or no interest to the ordinary person, nor perhaps even to the religious officials. Plato's story of Atlantis (*Kritias* 108e–109c, 113c–121c) purports to be derived from Egyptian records and based on events that took place 9,000 years earlier, but this is clearly a fictional figure, not based on real records. While hereditary priesthoods in Greece may have provided opportunities for record-keeping before the advent of writing, long-term record-keeping seems to have been a slow and late development in Greece.[21] The group to whom such extended records would be of increasing interest, however, was that of the astronomers. In the course of the Hellenistic period they gained access to very ancient records of celestial observations from Babylon. That the correction of the *octaetēris* took place in Greece in the fifth century, well before the Hellenistic period, is testament to Greek astronomers' deep interest in the problem. What drove them to seek to solve it escapes us still.

It is in order to eliminate the small difference of about a day-and-a-half per eight-year period that astronomers develop further lunisolar cycles. In arithmetical terms, the aim of these lunisolar cycles is to find as nearly as possible a whole number of lunar months which corresponds to a whole number of solar years. The best system devised in antiquity for practical purposes was the 19-year cycle. This is usually attributed to Meton, and so named after him, though his colleague in Athens, Euktemon, and others are credited with it by Geminos (*Introduction to Astronomy* 8.50–8).[22] The Babylonians also devised a 19-year lunisolar cycle some time earlier in the fifth century BC, but whether Meton invented his cycle independently of this eastern version, we cannot tell.[23]

According to Geminos:

> . . . as the *octaetēris* was found to be in all respects incorrect, the astronomers around Euktemon, Philippos and Kallippos constructed another period, that of 19 years. For they found by observation that in 19 years there were contained 6,940 days and 235 months, including the intercalary months, of which, in the 19 years, there are 7. And of the 235 months they made 110 hollow and 125 full, so that hollow and full months did not always follow one another alternately, but sometimes there would be two full months in succession. . . . As there are 235 months in the 19 years, they began by assuming each of the months to have 30 days; this gives 7,050 days. Thus, when all the months are taken at 30 days, the 7,050 days are in excess of the 6,940 days; the difference is <110 days>, and accordingly they make 110 months hollow in order to complete, in the 235 months, the 6,940 days of the 19-years period.
>
> (Geminos, *Introduction to Astronomy* 8.50–58)

So, over a period of 19 years there were 6,940 days or 235 months, including seven intercalary months. Of the 235 months, the Greeks made 110 'hollow' (in other words, of 29 days each), and the remaining 125 'full' (of 30 days each). The imbalance between 'full' and 'hollow' months means that they cannot alternate throughout the cycle, but sometimes there would be two 'full' months in succession. Geminos explains how the devisers of the cycle arrived at 110 'hollow' months: all 235 months are initially assigned 30 days each, which gives a total of 7,050 days to the 19-year period. This overshoots the sum of 6,940 days of 235 lunar months by 110 days, so 110 months must each have one day omitted through the cycle, and they become 29-day months. To ensure as even a distribution of this omission as possible, he says that the Greeks divided the 6,940 days by 110 to get a quotient of 63, so that the 110 days were removed at intervals of 63 days.

If the 19-year cycle is left to run unchanged, in four cycles, or 76 years, it gains a day against a solar calendar of 365¼ days: 6,940 × 4 = 27,760 days, but 365¼ × 76 = 27,759 days. Geminos (*Introduction to Astronomy* 8.59–60) tells us that the astronomer Kallippos therefore refined the 19-year cycle by running it over four periods and removing the extra day that had accumulated over that period (presumably by making a 'full' month 'hollow'). Those who accept Geminos' testimony about the omission of days in Meton's cycle are agreed that the omitted days should be every 64th one, rather than every 63rd, since omission every 63rd day does not work out correctly.[24]

An inscription from Miletos, perhaps connected with one of the two fragmentary stone *parapēgmata* excavated there, records two summer solstices, one in the archonship (the chief magistracy in Athens) held by Apseudes (and therefore in 432 BC), the other in the archonship of Polykles (i.e. in 109 BC).[25] The observation in 432 BC is dated according to the Athenian calendar on 13 Skirophoriōn. Ptolemy (*Almagest* 3.1) also records that Meton and Euktemon – he treats the two as collaborators in this enterprise – observed the summer solstice in the archonship of Apseudes in Athens, on a date which he gives in the Egyptian solar calendar as 21 Phamenoth. The Athenian version of the date, 13 Skirophoriōn, is significant, as it is the same as that which Meton used for the start of his new cycle, according to Diodorus Siculus:

> In the archonship of Apseudes in Athens . . . Meton, the son of Pausanias, distinguished in astronomy, published the so-called 19-year cycle, making the beginning from the 13th of the month Skirophoriōn in Athens.
>
> (Diodorus Siculus 12.36.1–2)

So the epoch for Meton's new cycle was the summer solstice in 432 BC, expressed as 13 Skirophoriōn according to the Athenian lunar month and 21 Phamenoth according to the Egyptian solar calendar. The Egyptian date allows us to assign the solstice observation, and hence the epoch of the Metonic cycle, to 22 June in our calendar.[26] Let us examine the Athenian calendar date.

The Athenian date of 13 Skirophoriōn is a date expressed in terms of the local festival calendar.[27] This was a lunar calendar, which regulated the celebration of religious festivals in Athens, indicating the specific days of specific months on which the festivals were to be held and sacrifices were to be made. In effect, it also provided a framework for the political calendar in the city, since there was a tendency to avoid holding political meetings on religious festival days. The twelve lunar months in Athens were: 1. Hekatombaiōn, 2. Metageitniōn, 3. Boēdromiōn, 4. Pyanepsiōn, 5. Maimaktēriōn, 6. Poseideōn, 7. Gamēliōn, 8. Anthestēriōn, 9. Elaphēboliōn, 10. Mounichiōn, 11. Thargēliōn, and 12. Skirophoriōn.

The names and order are secured by a variety of forms of evidence. There have survived, for example, a number of 'sacrificial calendars' from the districts (demes) of Attika. The earliest surviving specimen, dating probably to the 430s BC, is from the deme of Thorikos (*IG* I^3.256 bis, p.958).[28] The sacrifices to various gods in each month of the year are listed in succession, from Hekatombaiōn to Skirophoriōn (only the name of Metageitniōn has had to be restored).[29] The year started on 1 Hekatombaiōn, which occurred on the evening of the first sighting of the new moon's crescent following the summer solstice.[30] Another way of expressing this is that the last month of the year, Skirophoriōn, usually included the summer solstice.[31] This captures practice better, as we see with the work of Meton, who observed the solstice in mid-Skirophoriōn. In this regard New Year's Day in Athens parallels the Jewish Passover and Christian Easter in being a movable feast tied to both lunar and solar phenomena.

As we have seen, to maintain alignment with the seasons, a lunar calendar eventually requires the intercalation of a thirteenth month. It is usually thought that in Athens this was achieved by repeating the sixth month, Poseideōn, but the evidence is not so emphatic on this point, and indeed indicates that months 1, 2, 6, 7 and 8 could be repeated for intercalation. For the fifth century BC, the only inscriptional evidence of intercalation in Athens is for month 1, Hekatombaiōn, to be repeated if necessary. This is allowed for in the so-called First-Fruits Decree of the 430s (*IG* I^3 78).[32]

Much work has been done in recent years (unfortunately not all of it published) to suggest that the Athenians did in fact avail themselves of the 19-year Metonic cycle in their civil calendar.[33] An initial cycle of 19 years between 432 and 419 BC appears to have encountered difficulties, with the intercalation perhaps applied inconsistently. Thereafter, down to 331 BC, an 'ideal' cycle has been discerned, incorporating a regular pattern of 'ordinary' years (O) and leap years (L) in years 2, 5, 8, 10, 13, 16 and 18, creating the following sequence:[34]

O L O O L O O L O L O O L O O L O L O.

This idea has been picked up and carried further into the Hellenistic period, so that the Metonic cycle is now reasonably established as the organising principle for leap years at least to the mid-third century BC.[35] What happened from the mid-third century to the last quarter of the second century BC is not clear, but for the years between ca. 120 BC and ca. AD 180, Müller has suggested that a regular Metonic cycle was consistently in use, with intercalations in years 3, 6, 8, 11, 14, 17 and 19 of each cycle, creating the following alternative sequence:[36]

O O L O O L O L O O L O O L O O L O L.

The Metonic cycle begins with a solar phenomenon (the summer solstice), which is dated according to a lunar calendar (13 Skirophoriōn). This lunar date provides us with an age for the moon – nominally 13 days old, since 1 Skirophoriōn is by definition coincident with a new moon, observed or calculated – at the time of the summer solstice in 432 BC. The solstice on 13 Skirophoriōn in 432 BC is the date from which New Year's Day must then be reckoned. This day, 1 Hekatombaiōn, should occur with the coming new moon 18 days later, after the last day of Skirophoriōn (let us assume that the month is 'full' with 30 days, rather than 'hollow' with 29). What the Metonic cycle must then do is tell the user that by the time of the next summer solstice, one solar year later, the moon will be eleven days older (12 lunar months + 11 days = 365 days), and therefore that the lunar date will be 11 days more advanced, e.g. it will have moved from 13 Skirophoriōn to (13 + 11 days =) 24 Skirophoriōn. New Year's Day, 1 Hekatombaiōn, will therefore be measured from a moon that is older by 11 days than it was in the previous year.

To gain an idea of what this algorithm would mean to an Athenian, let us imagine what impact it would have on our calendar. A solar year is approximately 11 days longer than twelve lunar months. If beside our solar calendar of 365 days we were to run a lunar calendar consisting of only twelve lunar months and therefore of 354 days, and if our lunar New Year's Day in 1995 happened to fall on 1 January, when there also happened to be a new moon, Table 3.2 shows when lunar New Year's Day would fall over a period of 19 years from 1995 to 2013:

Table 3.2

Year	*New Year's Day*	*Year*	*New Year's Day*
1995	1 January	2004	13 September
1995	21 December	2005	2 September
1996	10 December	2006	22 August
1997	29 November	2007	11 August
1998	18 November	2008	31 July
1999	7 November	2009	20 July
2000	27 October	2010	9 July
2001	16 October	2011	28 June
2002	5 October	2012	17 June
2003	24 September	2013	6 June

In this period our imaginary New Year has run from midwinter (in northern hemisphere terms) in 1995 to midsummer by 2013, and it has more or less maintained alignment with the new moon (the occasional discrepancy is due to our not taking full account of the proper lengths of the solar and lunar years). Because of the shorter length of the lunar year, we have in this range one solar year, 1995, which contains two lunar New Years.

This, in fact, is the real-life situation for Islamic New Year (the first day of Muharram), which is not fixed to a point in the solar year, but runs back progressively through the seasons, as Table 3.3 demonstrates for the same period.

Table 3.3

1995	31 May	2005	10 February
1996	20 May	2006	31 January
1997	9 May	2007	20 January
1998	28 April	2008	9 January
1999	17 April	2008	29 December
2000	6 April	2009	18 December
2001	26 March	2010	8 December
2002	15 March	2011	27 November
2003	5 March	2012	15 November
2004	22 February	2013	5 November

In 1995 the first day of Muharram fell on 31 May. By 2013 it is projected to fall on 5 November. Again, because of the difference between the solar year and the 12 lunar months, in 2008 two New Year Days happen to occur.

Now let us see what happens if we attempt to put a brake on the lunar New Year's Day by inserting an extra month in seven years out of 19, as in the Metonic cycle. We can intercalate this month in solar years 2, 5, 8, 10, 13, 16 and 18, to see what effect the 'ideal' version of the cycle has (see Table 3.4).

So after 19 years, New Year's Day has tracked up and down a limited period between 26 December and 24 January, until it has returned to almost exactly the same calendar date we started at. (The apparent discrepancy of one day at the end is again due to the approximations we have adopted for the solar year and the lunar 'year'.) Throughout the full cycle, each New Year's Day has occurred at the same phase of the moon, in this case at new moon or very nearly so, but the new moon has not fallen on the same calendar date until we have reached the end of the 19-year cycle. The full cycle brings sun and moon back into the same calendrical and astronomical relationship every 19 years.

We saw with Muslim Muharram that it was possible to have two New Year's Days in the one solar year. In Table 3.4 we have ended up with three solar years bearing two lunar New Year's Days (1997, 2000 and 2008).

Table 3.4

Julian year	*Lunar New Year's Day*	*Difference in days*	*Year in cycle*
1995	1 January	0	1
1996	21 January	-11+ 30	2
1997	9 January	-11	3
1997	29 December	-11	4
1999	17 January	-11+ 30	5
2000	6 January	-11	6
2000	26 December	-11	7
2002	14 January	-11+ 30	8
2003	3 January	-11	9
2004	22 January	-11+ 30	10
2005	11 January	-11	11
2005	31 December	-11	12
2007	19 January	-11+ 30	13
2008	8 January	-11	14
2008	28 December	-11	15
2010	16 January	-11+ 30	16
2011	5 January	-11	17
2012	24 January	-11+ 30	18
2013	13 January	-11	19
2014	2 January	-11	1
2015	21 January	-11 + 30	2

Nonetheless, the number of lunar months over the whole 19-year period has remained at 235.

Our own Western New Year's Day does not move around like this, so initially it may be hard for us to appreciate what Meton was doing. But if we shift our thinking to a festival whose timing similarly must tie sun and moon together, then we can understand better not only the sophistication of the Metonic cycle, but also the need for it. Such festivals are Chinese New Year, Jewish Passover and Christian Easter, all of which must maintain a relationship with a given phase of the moon – a new moon for Chinese New Year, or a full moon for Passover and Easter.[37] By a process of intercalation, the lunar 'year' is regularly braked so that the solar year can catch up, and thus the lunisolar festival is kept to a limited period of weeks over a period of 19 years. After this period, it repeats the solar dates of the first cycle. To illustrate the point, the following table gives the dates of Passover for the period between 1995 and 2013 (see Table 3.5).

But that customary expectations for the timing of festivals could jar against new ways of marking time we know from the situation that arose when the Gregorian calendar was finally introduced into England in the eighteenth century; and it may be inferred from contemporary grumblings about missed festival days in the calendar in Meton's own time.

When the Gregorian calendar reform was legislated in Great Britain in 1752, 11 days were skipped in that year so as to make up for the discrepancy which had built up over time between the Roman Julian calendar and the sun. Catholic countries in Europe had made the drastic change almost two centuries earlier, from 1582, when 10 days had to be skipped. We owe to the nineteenth-century imagination the stories of riots and deaths in England over the introduction of the new calendar, but all the same the reform did cause confusion, especially in the religious sphere and the related economic world. Major and minor festival days, which signalled agricultural activities and linked into profitable market days in the towns,

Table 3.5

Year	*Passover*	*Year*	*Passover*
1994	27 March	2004	6 April
1995	15 April	2005	24 April
1996	4 April	2006	13 April
1997	22 April	2007	3 April
1998	11 April	2008	20 April
1999	1 April	2009	9 April
2000	20 April	2010	30 March
2001	8 April	2011	19 April
2002	28 March	2012	7 April
2003	17 April	2013	26 March

were thrown into confusion between the 'Old Style' and the 'New Style' calendars.[38]

Similarly, the complexity of the Metonic cycle may have been both the cause of some initial teething troubles, and, as a result, the trigger for a jibe by Aristophanes about missed festival days in the calendar. In his *Clouds*, first produced in 423 BC, the chorus of Clouds presents a series of complaints on behalf of the Moon:

> She says that she does other good turns, but that you do not observe the days correctly at all, but make them run up and down, so that she says the gods threaten her each time, whenever they are cheated of a meal and go home not having had the feast according to the reckoning of the days. And then whenever you should be sacrificing, you are torturing and judging, and often when we gods are observing a fast, when we mourn for Memnon or Sarpedon, you are pouring libations and laughing. For this reason Hyperbolos, having been chosen by lot this year to be the sacred remembrancer, then was deprived of his garland by us gods. For this way he will know better that one must observe the days of one's life according to the moon.
>
> (Aristophanes, *Clouds* 615–26)

The usual view of this complaint about mistimed feasts and fasts is that officials had recently been tampering with the lunar festival calendar so that it slipped out of sync with the moon's true phases. This interpretation seems to me misguided on several counts. To start with, it ignores the internal evidence within the play itself, at *Clouds* 16–18, which demonstrates the still viable link between the moon's 'twenties' – that is, the last third of the lunar month, when its days were numbered in the twenties – and the customary due date towards the end of a month for the interest on loans. This should mean that lunar phase and lunar month were still coordinated. Furthermore, external evidence for tampering with the festival calendar in the fifth century is very limited. There is more later, from the early Hellenistic period, which is anachronistically allowed to represent similar practices in Aristophanes' time.[39]

The misalignment which underlies the moon's complaint in Aristophanes' *Clouds* may instead be between the lunar festival calendar and other, sun- or star-based, time-schedules at the time.[40] The political calendar of Athens, on the one hand, shows signs of a short-lived dalliance with solar time-reckoning towards the end of the fifth century BC; and, on the other hand, the star-based almanac was being developed by Meton and his contemporary Euktemon in Aristophanes' own time. Columella, the Roman agricultural writer of the mid-first century AD, talks of the calendars (*fasti*) of Meton, Eudoxos (the fourth-century astronomer) and others having been adapted to public sacrifices (Columella, *On Agriculture* 9.14.12), a reference presumably to their almanacs being coordinated with the religious festival calendar. It is possible that Aristophanes was reacting to the difficulties encountered by public officials in making the festival calendar align with such seasonal, sun- or star-based 'calendars'. On the other hand, we may be witnessing here some of the initial hiccups upon the introduction of the Metonic cycle as the regulator of the leap years in the festival calendar. It seems that it was not until after 419 BC, following the first run of the cycle, that it settled down into a regular rhythm of ordinary and leap years. Either way, the complaint in the *Clouds* would be indicative of a shift from the moon to the sun as the fundamental overseer of the religious and civic year.

It was important for Meton that he have a mechanism for measuring the solar year, or (in other words more meaningful to him) of marking the regular occurrence of the summer solstice. How the cycle managed this is not clear from the description given by Geminos, but the fact that Meton and Euktemon also established almanacs, or *parapēgmata* as they were called, may provide a clue. A *parapēgma* keeps track of the solar year by noting various star phases (risings and settings at dawn and dusk), so it could have provided the means to keep track of the date of the solstice by a means independent of the wandering lunar calendar. This suggests that Meton's 19-year cycle was attached somehow to a *parapēgma*.[41]

A similar combination of Metonic cycle (probably with its refinement, the Kallippic, although this no longer appears to survive) and a *parapēgma* is found on the Antikythera Mechanism. We shall need to return to the question of what a *parapēgma* was in the context of timekeeping schedules, since one is embedded in the workings of the Mechanism, but let us first investigate a little more closely the matter of approximating the true solar year, as I have done in the above examples. This leads us into the presence of the zodiac and the Egyptian calendar on the same Mechanism.

The zodiac

To the casual observer, the sun appears to wheel daily overhead in a circle. As we have seen in the previous chapter, to anyone observing the sun over an extended period it becomes clear that the sun moves along the horizon in its risings and settings over the course of its seasonal cycle. We noted also that persistent observers who watch the sun at dawn and dusk may notice that its seasonal shift is set against the backcloth of the 'fixed' stars (so-called because they seem not to move relative to one another, unlike the sun, moon and planets), so that it traces a distinctive path, which is angled to the parallel paths of the rising and setting stars (Figure 1.8). This path, we saw, is called the zodiac, and it comprises twelve constellations known now by their Roman names (Aries, Taurus, Gemini, Cancer, Leo, Virgo, Libra, Scorpio, Sagittarius, Capricornus, Aquarius, Pisces.)

The sun's apparent passage across each of these constellations may be used as a measure of the seasonal year, with each constellation marking out a rough twelfth of the year. The Babylonians used this method of marking out the solar year, and the Greeks and Romans borrowed it from them, recognising in the twelve-part division a series of solar or zodiacal 'months'. Eventually, the full circuit of the zodiac was divided up into twelve equal divisions of 30° each, which were named after their resident constellation. The distinction between actual zodiacal constellations of varying size and artificial zodiacal signs of an even 30° of arc is not attested in extant Greek texts before the third century BC; Aratos, in that century, is the first author we know of to divide the ecliptic into twelve equal arcs, which are defined by the presence of the zodiacal constellations but are not explicitly named after them.[42]

The zodiacal stars are imbued with meaning all of their own. In antiquity the sun was regarded as one of a special class of stars called *planētes* in Greek, meaning the 'wanderers'. We call them planets. To the ancients they were gods, or their living agents.[43] The Greeks and Romans (and, significantly, the Babylonians before them) numbered among the planets Saturn, Jupiter, Mars, the sun, Venus, Mercury and the moon. Like the sun, these planets have paths which fall within the area of the zodiac. The zodiacal constellations therefore gained even more in significance as the 'home' of the planetary gods. Once the planets were seen as influencing human life on earth through their own special character, astrology was born.

Our interest here lies instead in the zodiac's function to mark time through the solar year, which so sophisticated an instrument as the Antikythera Mechanism still made use of. We shall defer till later in this chapter the issue of how the Mechanism utilises the zodiac in detail, since it involves the *parapēgma*, and that is more readily examined after we have investigated its use of other cycles related to the sun. Let us first examine the Egyptian calendar.

The Egyptian calendar

I mentioned just now the testimony of Columella that the 'calendars' – presumably the *parapēgmata* – of Meton and Eudoxos were adapted to religious festivals. When this occurred we do not know, but by about 300 BC a *parapēgma* had been not only linked to the Egyptian calendar but also provided with the dates of local religious festivals. This 'festival calendar' for the temple of Neith at Sais, southwest of Alexandria in the Nile Delta, survives nowadays as *P. Hibeh 27*.[44] It contains various types of information: when the sun enters each sign of the zodiac,[45] indications of when certain stars rise or set, measurements of the length of day and night (by water clock), days when festivals are due to take place, and weather forecasts. The following excerpt gives the readings for the Egyptian month of Mecheir (our early April to early May):

Mecheir 6 [The sun is] in Taurus. The Hyades set in the evening, the night is $11\frac{1}{2} + \frac{1}{10} + \frac{1}{30} + \frac{1}{90}$ hours, and the day is $12\frac{1}{3} + \frac{1}{45}$, and Hera burns.
And there is a change in the weather, and the south wind blows, but if it gets strong it burns up the produce of the land.

19 Lyra rises in the evening, the night is $11\frac{1}{3} + \frac{1}{15} + \frac{1}{45}$ hours, and the day $12\frac{1}{2} + \frac{1}{15} + \frac{1}{90}$ and there is an assembly at Sais for Athena, and the south wind blows, but if it gets strong it burns up the produce of the land.

2. . . . . rises in the evening, [the night is 11 . . . hours], and the day 12[.,] they observe [. . .]

27 Lyra sets in the evening, the night is $11\frac{1}{6} + \frac{1}{90}$ hours, and the day $12\frac{2}{3} + \frac{1}{10} + \frac{1}{30} + \frac{1}{45}$. Feast of Prometheus whom they call Iphthimis, and the south wind blows, but if it gets strong it burns up the produce of the land.

(*P. Hibeh* 27.66–87[46])

In effect, the astronomical observations, including the passage of the sun through the zodiac, have all been indexed against the Egyptian solar calendar, which is the fundamental organising principle in the schedule.[47] Herodotos (2.4) was aware of this very old calendar, taking the opportunity in his discussion of a Greek biennial intercalary system to praise the Egyptian calendar for keeping better pace with the seasons.[48] It was an administrative calendar, in which each year had exactly 365 days, divided into twelve months, each of 30 days, plus five extra days (*epagomenai* in Greek). The names of the months were: 1. Thōth, 2. Phaōphi, 3. Hathyr, 4. Choiach, 5. Tybi, 6. Mecheir, 7. Phamenōth, 8. Pharmouthi, 9. Pachōn, 10. Payni, 11. Epeiph, and 12. Mesorē. The 'epagomenal' days were tacked on at the end of the year after the month of Mesorē. The month drawn from the *Hibeh Papyrus* above is therefore the sixth in the year.

The overall length of the Egyptian year in this calendrical system may have been derived from long-term observations of natural phenomena, notably the regular, annual flood of the Nile, and the almost coincident dawn rising of the star Sothis (i.e. Sirius). The Egyptian names of their three seasons of the year reflect the country's dependence on the river: 'inundation', 'emergence' (of the fields from the flood waters), and 'dryness' (of the river before the next flood). Each season comprised four months: 'inundation' occurred from Thōth to Choiach; 'emergence' from Tybi to Pharmouthi; and 'dryness' from Pachōn to Mesorē. The summer flood of the Nile ('inundation') was agriculturally and calendrically the first season in Egypt, as it brought with it the necessary silt, into which grain seed could be sown in winter ('emergence'), to be harvested in spring ('dryness') (cf. Diodoros 1.11, 12, 16).

From a very early stage the annual flooding of the Nile was connected with the first sighting at dawn of the rising of the prominent star Sirius, called Sothis or the star of Isis by the Egyptians. Already by the time of the First Dynasty a text describes Sothis as 'Bringer of the New Year and of the Inundation'.[49] The same conception is embedded in the Ptolemaic Canopus Decree of 238 BC, which orders that:

> a public assembly be celebrated every year in the temples and throughout the whole country for King Ptolemy and Queen Berenike, the Benefactor Gods, on the day on which the star of Isis rises, which is considered in the holy books to be the New Year, and which is celebrated now in the ninth year on the first day of the month Payni, on which the Little Boubastia and the Great Boubastia are celebrated, the gathering in of the crops occurs and the rising of the river . . .
>
> (*OGIS* 56.35–8)

By this stage, however, the inherent weakness of the Egyptian calendar has become very obvious: no longer is the astral event occurring on 1 Thoth, but is now falling on 1 Payni, the tenth month of the traditional year. A year of 365 days is very close to the true solar year, but even in a person's lifetime its deficiencies would begin to show. Geminos explains the problem:

> The Egyptians have distinguished and calculated in a manner which is the opposite of the Greeks. For they do not observe that the years run according to the sun, nor the months and days according to the moon, but they have used a principle which is peculiar to them. They want the sacrifices to the gods to occur not at the same moment of the year but to pass through all seasons of the year, and the summer festival to occur in winter and autumn and spring as well.
>
> For they have the year of 365 days: they observe twelve 30-day months and five epagomenals. They do not add the extra quarter for the reason above, so that for them the festivals retrogress. For in four years they fall a day behind with respect to the sun, and in 40 years they will fall ten days behind with respect to the solar year, so that the festivals will also retrogress the same number of days, until they occur in the same seasons of the year. In 120 years the difference will be one month, both with respect to the solar year and with respect to the seasons of the year.
>
> (Geminos, *Introduction to Astronomy* 8.16–19)

Censorinus (*On the Birthday* 18.10) makes a similar point about the loss of approximately a day every four years. And it is not as if the Egyptians – or rather their Greek rulers – were not aware of the problem. Ptolemy III decreed that an extra day should be added every fourth year to correct the wandering year, thereby creating a leap-year system. The Canopus Decree itself records this intention:

> In the reign of Ptolemy [III] son of Ptolemy [II] and Arsinoe, the Brother-Sister Gods, in the ninth year, in the time of Apollonides son of Moschion, priest of Alexander and of the Brother-Sister Gods and of the Benefactor Gods, and of Menekrateia daughter of Philammonos, the basket-bearer of Arsinoe Philadelphos, on the 7th of the month Apellaios and the 17th of Tybi of the Egyptians; a decree . . . : so that the seasons also may run properly forever in accordance with the present state of the cosmos, and lest it happen that some of the public festivals, which are celebrated in winter, are ever celebrated in summer, since the star shifts one day every four years, while others, which are celebrated now in summer, are celebrated in winter, at the appropriate times hereafter, just as it has happened to be before, and would have been so now if the organisation of the year, from the 360 days and the five days which were deemed later to be intercalated, held good, *from the present time one day at the festival of the Benefactor Gods to be intercalated every four years after the five which are intercalated before the new year*, so that everyone may see that the correction and restoration of the previous deficiency in the organisation of the seasons and of the year and of the customs to do with the whole regulation of the heavenly sphere has happened through the Benefactor Gods.
>
> (*OGIS* 56.1–3, 40–6)

Generally in Egypt there is no evidence that anything was done to rectify the problem of the drifting calendar according to the decree.[50] As we have seen, at an early stage New Year's Day in the Egyptian calendar was equated with the day of Sirius' heliacal (pre-dawn) rising. This day should be 1 Thoth. But the calendar was a quarter-day short of representing the true solar year, and because of this lack of just a few hours each year, the calendar ran adrift of the seasons over a long period of time. Over time, 1 Thōth moved through every season of the year and so the actual day of Sirius' rising could not remain the marker for the start of the civil year. If we calculate this out, it took 1461 Egyptian years (the so-called Sothic cycle) before the start of the year could coincide again with the heliacal rising of Sirius. According to the Canopus Decree the rising of Sirius then fell on 1 Payni; this means that 1 Thōth occurred on 22 October. But because the rising of Sirius still coincided with the rising of the Nile and the start of the agricultural cycle, it was 1 Payni, rather than 1 Thōth, which was regarded as New Year's Day.

We can chart this shift of the calendar against the seasonal year for any given year. Table 3.6 demonstrates the situation in the years immediately after the passing of the Canopus Decree.

Table 3.6 Julian dates for 1 Thōth in 238–206 BC

New Year's Day	*238* BC	*234* BC	*230* BC
1 Thōth	22 Oct	21 Oct	20 Oct
	226 BC	*222* BC	*218* BC
1 Thōth	19 Oct	18 Oct	17 Oct
	214 BC	*210* BC	*206* BC
1 Thōth	16 Oct	15 Oct	14 Oct

The effect of the drift is of course more marked over longer periods of time, as Table 3.7 shows for the years 432 BC (when Meton 'observed' the summer solstice), 238 BC (the year of the attempted reform of the calendar under the Ptolemies), and 100 BC (around the supposed time of the Antikythera Mechanism).

In almost two centuries, between 432 and 238 BC, 1 Thōth has drifted from the equivalent of 9 December back to 22 October. In another 138 years by 100 BC, it has drifted back further still to 17 September. It continues this drift until the Roman takeover of Egypt in 30 BC, when 1 Thōth is at 29 August and then is fixed at that date in what becomes known as the Alexandrian calendar. A leap-year still had to be instituted, and it was, either in that same year or in 26 BC.[51] Until then the fixed length of the Egyptian year at 365 days proved too strongly embedded to correct, no doubt because of its religious associations. The Greek Ptolemies, who ruled Egypt in the Hellenistic period, saw the problem, sought to fix it, but failed. The Romans had less patience.

Table 3.7 Julian dates for Egyptian dates in 432 BC, 238 BC and 100 BC

Month	*432* BC	*238* BC	*100* BC
1 Thōth	9 Dec	22 Oct	17 Sep
1 Phaōphi	8 Jan	21 Nov	17 Oct
1 Hathyr	7 Feb	21 Dec	16 Nov
1 Choiach	9 Mar	20 Jan	16 Dec
1 Tybi	8 Apr	19 Feb	15 Jan
1 Mecheir	8 May	21 Mar	14 Feb
1 Phamenōth	7 Jun	20 Apr	16 Mar
1 Pharmouthi	7 Jul	20 May	15 Apr
1 Pachōn	6 Aug	19 Jun	15 May
1 Payni	5 Sep	19 Jul	14 Jun
1 Epeiph	5 Oct	18 Aug	14 Jul
1 Mesorē	4 Nov	17 Sep	13 Aug
epagomenal 1	4 Dec	17 Oct	12 Sep
epagomenal 2	5 Dec	18 Oct	13 Sep
epagomenal 3	6 Dec	19 Oct	14 Sep
epagomenal 4	7 Dec	20 Oct	15 Sep
epagomenal 5	8 Dec	21 Oct	16 Sep

The Romans' own calendar problems in the mid-first century BC were such – the calendar in the 50s BC alone had drifted 90 days out of sync with the seasons – that Julius Caesar took the opportunity as both *pontifex maximus* (the official in charge of the Roman calendar) and dictator to impose on the Roman state a drastic revision. In 46 BC he replaced completely the old Republican, quasi-lunar calendar, which comprised 355 days plus an intercalary month of 22 or 23 days, with a solar calendar of 365 days plus an extra day every four years. Significantly, he had Greek-Egyptian assistance in this project, notably from Sosigenes of Alexandria, who wrote three treatises on the subject (Pliny, *Natural History* 18.211–212). An initial hiccup in the calculation of when to add the necessary leap day caused the intercalation to occur every three years until 9 BC. Augustus finally brought matters back to order when he omitted leap days in three years (5 BC, 1 BC and AD 4), and restarted the Julian calendar properly from AD 5–8. A version of it was adopted in Egypt probably in 26 BC, so that a leap day was finally introduced into the Egyptian calendar, added as a sixth epagomenal day.[52]

Despite this amendment, because no allowance had to be made for leap years in calculations using the traditional Egyptian 365-day calendar, it was found to be extremely useful for long-term astronomical calculations, and so was adopted for dating observations by Greek astronomers (and was so used by astronomers down to the time of Copernicus).[53] In itself the Egyptian calendar's appearance on as advanced a scientific instrument as the Antikythera Mechanism therefore occasions no surprise. How it dealt with the leap day phenomenon has only just been elucidated, and this takes us to another dial, and another form of time-measurement.

The dial on the Mechanism formerly identified as providing the Kallippic Cycle has now been reclassified as presenting the four-year Olympiad cycle.[54] We have already seen that this cycle was probably governed by a longer *octaetēris* in the calendrical context. Various epochal eras were used by the Greeks and Romans.[55] The four-yearly periods of the Olympic Games formed the basis of the best-known era, that of the Olympiads, which started traditionally in 776 BC. Its invention is associated with Timaeus (ca. 350–260 BC) and Eratosthenes (ca. 285–194 BC).[56] As the Olympic year began in midsummer, it straddled the second half of one Julian year and the first half of the next, so that, for example, the third year of the sixth Olympiad (conventionally written as Ol. 6, 3) corresponds to the Julian years 754/3 BC. The other pan-Hellenic competitions could be referred to the Olympiad cycle, with the Olympic Games being assigned year 1. The Olympiad dial on the Mechanism had to be turned a quarter-turn each full year, so that after four years it completed the full Olympic cycle. On the dial, the Isthmian Games are listed under year 1 with the Olympic Games, but also in year 3, which they share with the Pythian Games. The Nemean Games come under years 2 and 4, and the Naa at Dodona in year 2 as well. Some other game or games were included in year 4, but the list is lost. The Olympiad dial also provided the means for recognising when the Egyptian dial, with its 365 days, had to be moved back a day every four years to take account of the leap year.

The *parapēgma*

I stated earlier that we do not yet know the Mechanism's purpose. Although currently recognised more as a planetarium or orrery, which mechanically models the relative positions and motions of the sun, moon and some or all of the known planets, it was originally identified as a 'calendar computer' by Price, and despite our current concerns with its sophisticated means of predicting lunar positions and eclipses, it still offers abundant means of marking time.[57] This it was able to do in terms of the Egyptian calendar on the front of the Mechanism, and of the Metonic lunisolar cycle of 19 years and perhaps of its refinement, the Kallippic cycle of 76 years, on the back of the Mechanism, where we also find the Olympiad four-year cycle, and the eclipse cycles.[58]

A further solar timing system was also used in the Mechanism: one of the front dials was a circular zodiac, concentric with the Egyptian calendar dial and set within it. The dial was divided into the twelve signs of the zodiac, creating effectively a second set of twelve months of the solar year alongside the Egyptian ones. Via letters of the Greek alphabet engraved against some of the days of the zodiacal months, this zodiacal cycle seems to have been keyed into yet another system, which was neither lunar nor strictly solar but sidereal. This system has traditionally been called a *parapēgma*, or 'star calendar', although both terms are loose in this context. This *parapēgma* was inscribed apparently on the front plate of the Mechanism, above and below the zodiac and Egyptian calendar dials. Fragments of the *parapēgma* are preserved embedded in the surviving plates, and more elements have been deciphered in recent years.

On the circular zodiac, there survive most of the month of *Parthenos* (Virgo), all of *Khēlai* (Libra), and the beginning of *Skorpios* (Scorpio). Of the names of the signs, only all of *Khēlai* and (we now know) *Skorpios* are there, but their presence as successive signs justifies the identification of *Parthenos* and the assumption of the rest of the zodiac. All are subdivided into what are taken to be degrees of arc, although they could as easily be days of the month (we do not have the full circle, which would resolve this issue). Price thought that he could discern the Greek letter *ōmega* inscribed above Virgo 18, then *alpha* above Libra 1, *bēta* above Libra 11, *gamma* above Libra 14, *delta* above Libra 16, and *epsilon* above Scorpio 1.[59] He assumed that there was only one alphabet's worth of letters attached to the zodiac, giving us a maximum of 24 readings in the associated *parapēgma*. New X-ray images have recently revealed further letters buried under corroded plates. We can now read beyond the *epsilon* at Scorpio 1, the letters *zēta* at Scorpio 5, *ēta* at Scorpio 17, *thēta* at Scorpio 22, and *iōta* at Sagittarius 1, as well as the name *SKORPIOS* itself. As there are now eight letters from a single alphabet incontrovertibly in just two signs – *alpha* to *thēta* in Libra and Scorpio – and as each sign for which we have the beginning currently has a letter at that point – *alpha* at Virgo 1, *epsilon* at Scorpio 1, and *iōta* at Sagittarius 1 – the present investigators of the Mechanism, the Antikythera Mechanism Research Group, have suggested that two full alphabets' worth of letters might have been inscribed, with the second set finishing with *ōmega* at Virgo 18, and that there was a letter at the first degree of every zodiacal sign.

It was also taken for granted by Price that the letters inscribed around the zodiac were linked directly to an inscription on the front face of the Mechanism. He transcribed the best preserved section (Fragment C) of this inscription, working not only from the X-rays he had had made but also from photographs taken early in the twentieth century, which preserve parts now lost to further corrosion. Recent discoveries do not materially overturn his interpretation, but a more accurate translation would run thus:

[K	]e[vening]
Λ	The Hyad[es set in the e]vening
[M]	Taur[us begins to r]ise
[N	L]yra [rises in the e]vening
[Ξ]	The Pleiad rises in the mornin[g
O	The Hyad rises in the morning
Π	Gemini begins to rise
P	Eagle rises in the even[ing
Σ	Arcturus sets in the morn[ing][60]

We have a set of alphabetic letters in sequence, beside which is attached a series of observations (for want of a better word) of star-rise and star-set phenomena.[61] This content identifies the inscription as part of a *parapēgma*, or what we may loosely term a 'star calendar'. We shall return to the question of what constituted a *parapēgma*, and what it did, but for the moment let us stay with the inscriptions on the Mechanism. There is, unfortunately, no overlap yet between these letters and those that we have seen inscribed on the zodiac dial – one set stops at *iōta* in Sagittarius, the other starts with *kappa* (probably in Aries)[62] and therefore any correlation between these two elements of the Mechanism is admittedly hypothetical.[63] In addition, a recent reinterpretation of what is left of a smaller, less well-preserved section (Fragment 22) would indicate the presence of a second set of alphabetic letters. This fragment also gives M, N, Ξ, O and Π in sequence, but attached this time to different stars (*Ōriōn* [Orion] with N, *Kyōn* [Sirius] with Ξ, and *Aetos* [Aquila] with O). Two alphabets' worth of letters in the 'star calendar' coincides with two sets dotted along the zodiac dial, but we still lack any overlap between the dial and the 'star calendar'.

To understand what all this might signify, we need to backtrack and look briefly at what we know about Greek *parapēgmata* in general. In literature, Homer's *Iliad* and *Odyssey* display awareness of the use of both the sun and the moon as means of reckoning time, the sun for the seasonal year, and the moon for other measures, such as the length of a pregnancy, a form of time reckoning which has continued to the present day. No calendar as such is mentioned, nor are months named, but instead the risings and settings of a few significant stars are used to signal certain periods in the seasonal year. As we saw in chapter 1, the stars observed at sunrise and sunset provided a useful sequence of first and last visible risings and settings. These systematic observations of the stars are developed further in Hesiod's *Works and Days*, where the poet gives an account of the activities required of the farmer in the agricultural year. Hesiod provides ten observations of the rising, setting or (in one instance) culmination of five stars or constellations, which help to distinguish four seasons. Reiche pointed out how sophisticated Hesiod's schedule of star observations could be, if excerpted and arranged diachronically, for the prime agricultural activities of ploughing, sowing and harvesting.[64] While ten observations of the risings, settings, or culmination of five stars or star groups may seem a very small number over a year, Reiche rightly pointed out that Hesiod's economical set of data still provides a functional safety net of observations over the crucial parts of the agricultural year, from sowing to reaping.[65]

Clearly it was not Hesiod's principal aim to write about astronomy. But embedded in his poem are indications of an underlying familiarity on the poet's part, and presumably on his readers' part too, with a body of astronomical material, which did not need to be explained and which was quite likely much greater than what Hesiod has introduced into this one poem. Stars are identified, and astronomical times of year are mentioned, in the *Works and Days*, implying processes of comprehension, conceptualisation and categorisation that have already taken place at some indeterminate period in the past. Hesiod's data may have derived from a dedicated astronomical poem – an *Astronomica* was attributed to him – and it would be interesting to know how this compared with Egyptian and Babylonian texts, which are earlier, more extensive, and more systematised than what Hesiod gives us.[66]

Star-lore remained in use throughout later periods, providing historians with better temporal fixes for their narratives than the relatively discordant, local state calendars. Thucydides, for instance, in the late fifth century BC recommended the use of summers and winters to mark the passage of time from one year to the next in the Peloponnesian War (5.20.1–2):

> This treaty was made as winter was ending, in the spring, immediately after the city Dionysia, just ten years and a few days over having elapsed from when the invasion of Attika and the beginning of this war first took place. This must be considered according to the

> periods of time, and not by trusting the counting of the names everywhere of those who either from holding office or from some other honour act as markers for past events. For accuracy is not possible, where something may have occurred while they were at the beginning of office, or in the middle, or however it happened to be. But by counting according to summers and winters, as this is written, it will be found that, each having the force of a half of a year, there were ten summers and as many winters in this first war.
>
> (Thucydides 5.20.1–2)

Elsewhere (2.78.2; 7.16.2) he used the first visible dawn rising of the star Arcturus and the winter solstice as temporal markers.

In much the same period, Greek medical writers similarly used the equinoxes and the risings and settings of stars to refer to different seasons of the year. For example, one writer associated with the Hippokratic school (but not Hippokrates himself) tells us:

> Anyone who reflects on and considers these things may foresee most of what will result from the changes. One should especially beware of the greatest changes of the seasons, and neither give medicine willingly, nor cauterise the belly, nor cut until ten or more days have past. These are the greatest and most dangerous, the two solstices, and especially the summer; and the two equinoxes are also so considered, but especially the autumnal. One should also beware of the rising of the stars, especially of the Dog, then of Arcturus, and then the setting of the Pleiades; for illnesses reach their crises especially in those days, and some are fatal, some cease, while all others change to another form and another state. So it is with regard to these matters.
>
> ([Hippokrates], *On Airs, Waters, and Places* 11)

In another medical text, winter is given as lasting from the setting of the Pleiades to the spring equinox, spring from the equinox to the rising of the Pleiades, summer from then until the rising of Arcturus, and autumn from the rising of Arcturus to the setting of the Pleiades ([Hippokrates], *On Regimen* 3.68.2).

That such astronomical information was popularly comprehensible at the time is demonstrated by the writer of the play *Rhesos* (once thought to be Euripides), who demonstrates a similar familiarity with the use of stars as time-keepers when he has his chorus of Trojan soldiers call out:

> Whose watch is it? Who is taking mine? The first signs are setting and the seven-pathed Pleiades are on high; the Eagle flies in the middle of the sky. Wake up! Why are you delaying? Get out of bed

to the watch! Do you not see the light of the moon? Dawn is near, dawn is coming and this star is one of the vanguard!

([Euripides], *Rhesos* 527–36)

From the late fifth century BC onwards Greek astronomers formalised data on star-rise and star-set into some form of almanacs or lists called *parapēgmata*. These survive today through archaeological excavations as stone tablets inscribed with day-by-day entries for the appearance or disappearance of stars.[67] Many of the leading astronomers of antiquity played a role in the development of the *parapēgmata* – the list includes Meton, Euktemon, Eudoxos, Kallippos and Ptolemy – and they continued in use until the Medieval period in Europe, being superseded only on the invention of the mechanical clock, which permitted the measuring of time regardless of the weather conditions.[68] In literary form, these *parapēgmata* were combined and published either in their own right (as in Geminos' *Introduction to Astronomy*, or in Ptolemy's *Phaseis*), or subsumed into agricultural 'handbooks' (such as Columella's *On Agriculture*, and Varro's *On Farming*). Star knowledge of this kind pervades every aspect of Greek and Roman literature: it can be found in all the major authors, from Aiskhylos to Euripides and Aristophanes, through Aratos to Plautus, Vergil and Ovid and beyond.[69] Julius Caesar himself was credited with a *parapēgma*, which survives in later quotations (Pliny, *Natural History* 1, 18.214).

As large-scale stone inscriptions set up in cities, *parapēgmata* survive in fragments from across the Hellenistic and Roman Mediterranean, most notably the specimens from Miletos of the late second – early first centuries BC.[70] The term *parapēgma* was known primarily from its attachment to the literary list of star-rises and star-sets, which is appended to the end of Geminos' *Introduction to Astronomy*.[71] In what sense anything in this list was 'stuck beside', as the word *parapēgma* implies, was not obvious, until the discovery of physical remains of similar star-lists. These stone tablets gave daily notifications of star events (as well as other information, particularly some weather 'predictions'), beside which were holes for the insertion of a peg. The findspots of actual *parapēgmata* in public, civic contexts also reaffirm further the public character of such lists: *parapēgmata*, like decrees, served a civic purpose well beyond the interests of the astronomers.[72]

The following excerpt from the earlier of the two *parapēgmata* from Miletos (MI, 109/8 BC) provides a useful guide to the type. The mark ○ here indicates peg-holes in the original; words in angled brackets have been restored; modern constellation equivalents are in square brackets:[73]

○ The Sun in the Water-Pourer [Aquarius]

○ <The Lion> [Leo] begins setting in the morning and the Lyre [Lyra] sets

○ ○
○ The Bird [Cygnus] begins setting at nightfall
○ ○ ○ ○ ○ ○ ○ ○ ○
○ Andromeda [Andromeda] begins to rise at dawn
○ ○
○ The Water-Pourer [Aquarius] is in the middle of its rising
○ The Horse [Pegasus] begins to rise in the morning
○
○ The whole Centaur [Centaurus] sets in the morning
○ The whole Hydra [Hydra] sets in the morning
○ The Ketos [Cetus? Pisces?] begins to set in the evening
○ The Arrow [Sagitta] sets, a season of continuous west winds
○ ○ ○ ○
○ The whole Bird [Cygnus] sets in the evening
[○ <Arktouros [Arcturus] rises> in the evening]

This version links the observations to the artificial signs of the zodiac, a system that we find also in the Geminos compilation. Such a systematisation of a *parapēgma* must have occurred by ca. 300 BC, to judge from its appearance in the festival calendar of that date from Sais in Egypt, which we have already examined. In this we find a *parapēgma*, resembling in some of its details what we know about that of Eudoxos, structured according to the native Egyptian calendar, and incorporating notices of the movement of the sun into each zodiacal sign. To give a further example, here is the Egyptian month of Tybi (approximately our March):

Tybi [5]	[The sun is] in Aries
20	Spring equinox, the night is 12 hours and the day is 12 hours, and the feast of Phitorois
27	Pleiades set in the evening, the night is $11\frac{2}{3} + \frac{1}{6} + \frac{1}{90}$ hours, the day $12\frac{1}{10} + \frac{1}{30} + \frac{1}{45}$

(*P. Hibeh* 27. 62–66[74])

The later of the two Milesian *parapēgmata* (MII, 89/88 BC) attributes the star and weather observations to various astronomers:[75]

. . . evening . . .
according to Euktemon ○
○ The Goat [Capella] sets acronychally according to both Philippos and the Egyptians
○ The Goat [Capella] sets in the evening according to the Indians' Kallaneus ○
○ The Eagle [Aquila] rises in the evening according to Euktemon

○ Arktouros [Arcturus] sets at dawn and there is a change in the weather according to Euktemon. On this day the Eagle rises in the evening also according to Philippos

The compilation-*parapēgma* in Geminos combines both of these systems, presenting the data organised by zodiacal months, and including attributions to astronomers from the fifth to the third century BC.[76]

If we were to extract those observations attributed only to Euktemon, for example, we would have the following for the same 'month' of Aquarius that the earlier Miletos *parapēgma* presented:[77]

[The sun is in Aquarius] (24 January–23 February)
○ ○
○ Lyra sets in the evening. Rainy weather.
○ ○ ○ ○ ○ ○ ○ ○ ○ ○ ○ ○ ○
○ Time of the beginning of the west wind.
○ ○ ○ ○ ○ ○ ○
○ Sagitta sets in the evening. Very stormy weather.
○ ○ ○ ○ ○ ○

Overall, the Geminos text of Euktemon gives a greatly increased number of observations over Hesiod's – 42 observations of 15 stars or star groups – together with notices of the solstices and equinoxes. The solstices and equinoxes quarter the year, and we can even find in Euktemon's list evidence of 'mid-quarter' days, the points in between each successive pair of solstice and equinox. These mid-points serve as markers for significant farming activity across the end of one season and the start of another in various cultures.[78]

We may imagine such lists of observations being drawn originally from lists arranged by simple day-counts between the observations. This system is explicitly used by Hesiod, and later in a full *parapēgma* which survives only in a fifteenth- century manuscript in Vienna, and the observations in which were thought by Rehm to be basically those of Euktemon, on the basis of the stars enumerated and the language used to describe the phenomena. The same period of the year that we have been illustrating so far is expressed thus:[79]

From the setting of the Eagle [Aquila] to the Dog [Sirius] 4 days. The Etesian winds begin to blow.
From the appearance of the Dog [Sirius] to the setting of the Lyre [Lyra] and the rising of the Horse [Pegasus] 13 days.
From the setting of the Lyre [Lyra] and the rising of the Horse [Pegasus] to the appearance of Protrygeter [Vindemiatrix] and the rising of Arcturus and the setting of the Arrow [Sagitta] to the appearance of Arcturus and the rising of the Goat [Capella] 10 days.
(cod. Vind. Gr. philos. 108, fol. 282v, 283r)

Whether this *parapēgma* is based on Euktemon's or not,[80] it nevertheless probably represents the format by which star and weather observations were remembered and preserved over generations, until the lists were formalised in stone in the fifth century, and then, from the early third century onwards, organised into zodiacal months. Even after this reorganisation, this day-count system persisted, to judge from a curious *parapēgma* preserved in the so-called *Ars Eudoxi* (P. Par. 1), a papyrus dating to ca. 190 BC, but using Eudoxan astronomical data in a manner similar to the *Hibeh Papyrus* and therefore perhaps dating, like it, to ca. 300 BC.[81]

How the observations were made and with what instrumentation beyond the naked eye we do not know. Meton's teacher Phaeinos, a metic (resident foreigner) in Athens, observed the solstices from Mount Lykabettos in Athens ([Theophrastos], *On Weather-Signs* 4), while Meton himself set up *stēlai* and recorded the solstices (Aelian, *Miscellany* 10.7). Whether Meton's *hēliotropion* on the Pnyx in Athens (scholion to Aristophanes, *Birds* 997) is connected with this solstitial activity is moot. The name of the instrument suggests that it was connected with a solstice in some fashion, and it is remarkable, as we saw in the previous chapter, that the rising of the summer solstice sun, when seen from the Pnyx, occurs at or near the peak of Lykabettos (Figures 2.2, 2.3), so natural features may have assisted in the topographical definition of the solstice.

Others were helped by the *parapēgmata* to organise their activities in time with the seasonal year. Columella refers explicitly to the star calendars of Meton and Eudoxos being adapted to public sacrifices:

> Indeed, in this rural instruction I am now following the calendars of Eudoxus and Meton and the old astronomers, which are adapted to the public sacrifices, because that old view, understood by farmers, is better known, and, on the other hand, the subtlety of Hipparchus is not necessary, as they say, for the duller learning of rustics.
>
> (Columella, *On Agriculture* 9.14.12)

Hibeh Papyrus 27 shows how this assimilation could look, with its listing of local feast days alongside astronomical and meteorological data.

Certainly *parapēgmata* were associated with forecasting, although usually of weather rather than other events. Astrometeorology had a long history throughout antiquity,[82] and indeed it has been proposed that before the first century BC the Greeks were interested not so much in measuring time *per se* as in observing the orderly sequences of 'omen events' such as star-rise and star-set, equinoxes and solstices, on which the sequences of agriculture and religion relied.[83] This argument is attractive, but some caution is warranted: 'omen-events' need careful definition, if we are not to do an injustice to the *parapēgmata*, some of which (particularly the early Greek ones) expend much more space on pure fixed-star phase prediction than they do on the meteorological forecasts that might be read as a causal result of those star phases. The cultural context of astrometeorology in both the Near East and Greek worlds is now being given serious attention.[84]

This survey does not exhaust by any means the varied forms and functions of *parapēgmata* in the Greek and Roman worlds, but it provides sufficient context to appreciate the *parapēgma* on the Antikythera Mechanism.[85] Price was aware of the *parapēgma* attached to Geminos, and used the data attributed in it to Eudoxos to try to reconstruct the rest of the Mechanism's *parapēgma*, combining the data from both the zodiacal dial and the fragment of the star-list. This meant organising Fragment 22's star-list into the same zodiacal month structure as that of the dial, and deciding on the basis of Eudoxos' *parapēgma* which months were surviving in the Mechanism. The result (in slightly expanded form for ease of comprehension) was as follows:

Α	Libra 1	Autumn equinox
Β	Libra 8	Pleiades evening rise
Γ	Libra 22	Hyades evening rise?
Δ	Scorpio 8	Arcturus evening rise?
Ε	Scorpio 19	Pleiades morning set
Ζ	Scorpio 29	Hyades morning set
Η	Sagittarius 12 *or* 16	Sirius morning set *or* evening rise
Θ	Sagittarius 12 *or* Capricorn 18	Altair morning rise *or* evening set
Ι	Pisces 4 *or* Aries 1	Arcturus evening rise *or* Spring equinox
Κ	Aries 13	Pleiades evening set
Λ	Aries 21	Hyades evening set
Μ	Taurus rises	
Ν	Aries 27	Vega evening rise
Ξ	Taurus 22	Pleiades morning rise
Ο	Gemini 5	Hyades morning rise
Π	Gemini rises	
Ρ	Gemini 7	Altair evening rise
Σ	Gemini 13	Arcturus morning set
Τ	Cancer 1	Summer solstice
Υ	Cancer 27	Sirius morning rise
Φ	Leo 5	Altair morning set
Χ		?
Ψ		?
Ω	Virgo 19	Arcturus morning rise

(After Price 1974: 46, Table 4)

Price acknowledged that this was a very tentative reconstruction. Clearly it assumes a single alphabet's worth of letters for the whole *parapēgma*, an assumption that is now difficult, if not impossible, to maintain. In addition, Lehoux has reviewed this theoretical reconstruction in the light of all other *parapēgmata*, and has concluded that Price's attempt was misguided because of the poor fit with what the Geminos list preserves of Eudoxos' *parapēgma*.[86] This criticism deserves more analysis. It is true that the Mechanism omits from the Geminos version of Eudoxos' *parapēgma* Orion at Taurus 1, Sirius at Taurus 2, rain at Taurus 7, Capella (Goat) at Taurus 9, and Skorpios at Taurus 11 and 21. But the Mechanism's *parapēgma* is a perfect fit with Eudoxos for the month of Gemini with no omissions, and overall six of the Mechanism's observations are in Eudoxos. On the other hand, we may note that the Mechanism's *parapēgma* fits at least as well with Euktemon, again with six 'hits', omitting only Sirius at Taurus 2 (but covering Lyra that same day), and Capella (Goat) at Taurus 8, and misplacing Hyades at Taurus 32. Proportionally, then, given the smaller number of observations in Euktemon for this time of year, his *parapēgma* provides a better match than does Eudoxos'. This analysis seems to be supported by Ptolemy's *Phaseis*, a much later literary *parapēgma* of only first and second magnitude stars, which gives five matches for *klima* 14 (the latitude of Alexandria and Cyrene), whereas *klima* 14½ (the latitude of Rhodes) delivers only four – something that might speak against Eudoxos as the source for the Mechanism's *parapēgma*, since Rhodes was where he worked. Interestingly, the (Euktemonian?) Vienna manuscript gives six matches too, but the (Eudoxan?) *Hibeh Papyrus* only four.

Lehoux then turned his attention to the section preserved on the dial with the zodiacal signs. Using the autumnal equinox, which is marked by A at Libra 1 on the Mechanism, as a starting point, he compared the day-differences on the Mechanism's zodiac dial to day-differences measured from the equinox on other *parapēgmata*. He found closer correspondence to stellar phases in the *parapēgmata* of Columella, Ptolemy (for *klimata* 14 hours and 14 ½ hours in length), and the Late Antique Parisinus gr. 2419. None, he noted, is perfect, emphasising again how unique in the present record is the *parapēgma* on the Antikythera Mechanism. But this analysis is problematic too: the Mechanism omits many observations of other stars in Columella and Ptolemy over this same time period. Just for the period represented by Libra on the Mechanism, for example, we have only three observations in its *parapēgma* (Libra 1, the autumn equinox; Libra 8, the evening rise of the Pleiades; Libra 22, the evening rise of the Hyades?), while Columella provides the following lengthier list (I have translated his Roman dates to modern ones, and inserted the zodiacal dates, which he omits):

24, 25, 26 September (Libra 1): the autumn equinox signifies rain.
27 September (Libra 4): Haedi (the Kids in Auriga) rise, Favonius (the West Wind), sometimes Auster (the South Wind) with rain.

28 September (Libra 5): Virgo stops rising, signals stormy weather.
1 and 2 October (Libra 8 and 9): Signals stormy weather occasionally.
4 October (Libra 11): Auriga sets in the morning, Virgo stops setting, signals stormy weather sometimes.
5 October (Libra 12): Corona begins to rise, signals stormy weather.
6 October (Libra 13): Haedi rise in the evening, the middle of Aries sets, Aquilo (the North Wind).
8 October (Libra 15): the bright star of Corona rises.
10 October (Libra 17): Vergiliae (the Pleiades) rise in the evening, Favonius (the West Wind) and occasionally Africus (the Southwest Wind) with rain.
13 and 14 October (Libra 20 and 21): All of Corona rises in the morning, wintry Auster (the South Wind) and sometimes rain.
15 October (Libra 22) and on the following two days stormy weather occasionally, sometimes only wet with dew, Iugulae (all or part of Orion) rise in the evening.

(Columella, *On Agriculture* 11.66–76)

If we ignore the meteorological data, we find in the short passage from Columella the extra observations of the Kids in Auriga, Virgo, all of Auriga, part and all of Corona, Aries, and part or all of Orion.

We also now have more alphabetic letters available on the dial than Price could see.[87] Ongoing study of the Mechanism may help to clarify the form of its *parapēgma*. But can we deduce its purpose in the context of the instrument as a whole?

The Antikythera Mechanism as a 'time machine'

Two hundred or more years after the construction of the Mechanism, in the *Almagest* Ptolemy would use a combined system of the Egyptian calendar and the Kallippic cycle,[88] while in his *Phaseis* he also calibrated a composite *parapēgma* derived from various sources against the Egyptian calendar. In the Antikythera Mechanism we find all three systems of time measurement available.

Indeed, correlations with a further system would also be possible, namely with the local, civil, lunar calendars of the Greek world. I have suggested myself that this would have to be on an *ad hoc* basis, because there appeared to be no explicit reference to any civil system on the Mechanism.[89] But we now know that in fact a civil system was used on the Mechanism.[90] From fragmentary scraps of names around the Metonic cycle dial the full names of all twelve civil months can be reconstructed. It can also be determined which was the first month of this calendar's year: at each entry for it there is a symbol for 'year' and a number representing its place in the Metonic 19-year cycle. So the months can be organised into their proper calendrical sequence as follows: 1. Phoinikaios. 2. Kraneios. 3. Lānotropios. 4. Māchaneus. 5. Dōdekateus. 6. Eukleios. 7. Artemisios. 8. Psydreis. 9. Gameilios. 10. Agrianios. 11. Panāmos. 12. Apellaios.

These names are basically Dorian Greek in form, but, given the patchy state of our knowledge of most regional Greek calendars, it is difficult to define the provenance of the calendar any better. The best parallel lies in a regional calendar, recently reconstructed from fragmentary calendars in southern Illyria, Epiros, and Kerkyra (Corfu), in northwest Greece. In this same region Dodona is situated, where the four-yearly games, the Naa, were held, as we have seen already in relation to the Olympiad dial on the Mechanism. All but one of the Mechanism's months are paralleled in this calendar.[91] The odd one out, Lānotropios on the Mechanism, is still close in form to the twelfth known from other evidence, Ālotropios (or Hāliotropios). Nonetheless, this parallelism does not mean that the Mechanism came from this region. The Greek cities in the Epirote region which shared this calendar ultimately derived from Corinth, so the calendar might derive from there too, or from another Corinthian colony. Unfortunately, Corinth's own calendar is still barely known: we have only the months Phoinik[aios] and Panēmos from it.[92]

A Corinthian colony like Syracuse, rich in the history of astronomy (it was the home of Arkhimedes, the great scientist of the third century BC), may be as likely a candidate for the provenance of the Mechanism's calendar as Corinth itself.[93] But this is of no real assistance, since we know as little about that calendar: Plutarch (*Nikias* 28) equates Syracusan Karneios (cognate to the Mechanism's Kraneios) with Athenian Metageitniōn, which is the second month in Athens, beginning with the second new moon after the summer solstice, around mid-August; and an inscription from Magnesia on the Maeander in Turkey refers to the month Apollōnios at Syracuse.[94]

Much more problematic is the apparent testimony of the speech *On the Crown* by the Athenian orator Demosthenes, used by the Antikythera Mechanism Research Group to support its analysis of the Mechanism's calendar.[95] The text of this oration purports to quote a letter from Philip of Macedon, in which Corinthian Panēmos is equated with the Athenian month Boēdromiōn and the Macedonian month Lōios. Yet this 'testimony' is normally regarded as spurious nowadays on a number of grounds, a product of the later transmission of the speech (and others like it) from the late Hellenistic period onwards, when the documents mentioned by Demosthenes – decrees, laws, diplomatic letters – were imaginatively recreated, sometimes as a school exercise in rhetoric, on the basis of the speech itself.[96] This 'letter of Philip' would have Corinthian Panēmos correspond with the third month of Athens' calendar, Boēdromiōn. Yet in the Mechanism's calendar, while Kraneios is the second month and may plausibly be coincident with Athenian Metageitniōn (on the basis of Syracuse's Karneios), Panāmos would not be third, and hence coincident with Athenian Boēdromiōn (like Corinth's Panēmos), but eleventh. This is a salutary warning of the slippery nature of homonymous months in Greek calendars, even in regional calendars that might be expected to share a common ancestry.[97]

Indeed, using Panāmos at all to anchor the Mechanism's calendar is fraught with difficulty. Pritchett long ago pointed out that Panēmos/Panāmos was the most common of all the months in the surviving Greek calendars, and yet was situated in the year anywhere from April to October in our terms, depending on which city one was in.[98] Trümpy's more recent study of the regional calendars of Greece suggests that the month is the second-most common, after Artemisiōn/Artemisios, but nonetheless she confirms a wide range for Panēmos/Panāmos through the year, e.g. Panāmos in Epidauros corresponds to Athenian Thargēliōn (May/June); in Megara (as reconstructed by Trümpy) to Athenian Mounichiōn (April/May) or Thargēliōn (May/June); in Rhodes and Thessaly to Athenian Hekatombaiōn (July/August); and in Boiotia to Athenian Metageitniōn (August/September).[99]

This unstable situation stands in contrast with the regular placement of the month Artemisiōn/Artemisios in springtime, no doubt because of its association with the cult of Artemis. Panēmos/Panāmos lacks any such cultic association.[100] Cabanes argues that the Epirote month of Artemisios fell in February/March, just before the spring equinox, and that the New Year began with this month. It seems, however, that the Mechanism's year began instead with Phoinikaios. Nevertheless, we may trust in the springtime association of Artemisios and see how the Mechanism's year pans out. Following Cabanes' placement of Artemisios in February/March, Phoinikaios would signal the New Year at the equivalent of Metageitniōn, around August/September, just before the autumn equinox. This would be three months before New Year proposed in the Antikythera Mechanism Research Group's reconstruction.[101] Reality presumably resides at one of these extremes or somewhere between them. Let us tabulate these results, including the calendar from Athens as a recognisable guide, and for good measure let us add a calendar related to the Athenian one for comparison, that of Delos (see Table 3.8).

In the table, two versions of the Mechanism's year are presented: (1) is that implied by the Antikythera Mechanism Research Group's publication of the calendar, while (2) represents the effect of following Cabanes' placement of Artemisios just before the spring equinox. The first month of the year in each calendar is underlined. In Athens the year began after the summer solstice. Delos, despite having close political ties with Athens and matching several months in its own calendar, began its year after the winter solstice. The Mechanism's year would begin between the two, in late summer or late autumn.

Table 3.8

Athens	*Delos*	*Antikythera Mechanism (1)*	*Antikythera Mechanism (2)*
Hekatombaiōn	Hekatombaiōn	Gameilios	Apellaios
Metageitniōn	Metageitniōn	Agrianios	Phoinikaios
Boēdromiōn	Bouphoniōn	Panāmos	Kraneios
Pyanepsiōn	Apatouriōn	Apellaios	Lānotropios
Maimaktēriōn	Arēsiōn	Phoinikaios	Māchaneus
Poseideōn	Posideōn	Kraneios	Dōdekateus
Gamēliōn	Lēnaiōn	Lānotropios	Eukleios
Anthestēriōn	Hieros	Māchaneus	Artemisios
Elaphēboliōn	Galaxiōn	Dōdekateus	Psydreis
Mounichiōn	Artemisiōn	Eukleios	Gameilios
Thargēliōn	Thargēliōn	Artemisios	Agrianios
Skirophoriōn	Panēmos	Psydreis	Panāmos

Two independent studies have recently demonstrated that even in the period after the construction of the Mechanism, in the reign of Augustus, Greek civil calendars were probably still aligned with the moon, with the months being synchronised with the lunar phases.[102] Furthermore, from the late fifth century BC into the Hellenistic period, it also appears much more likely that the sequence of ordinary and leap years in the civil calendar of Athens was regulated by the Metonic cycle.[103] The Mechanism now makes it plain that a similar system certainly operated elsewhere, with the Metonic cycle probably serving as the regulator for a civil, lunar calendar.

On the Mechanism it looks likely that the system of intercalation allowed for an extra month in years 1, 3, 6, 9, 11, 14 and 17.[104] This differs from the intercalary systems identified epigraphically in Athens for the period from the late fifth century BC to the mid-third century BC – with intercalary months in years 2, 5, 8, 10, 13, 16 and 18 – or from about 120 BC onwards – with intercalary months in years 3, 6, 8, 11, 14, 17 and 19. It may be that even Athens did not maintain a uniform system over time in its use of the cycle. Certainly there is no reason to think that other Greek cities felt obliged to use Athens' system(s), and the Antikythera Mechanism indicates that another intercalary system was in operation that was at variance with that of Athens. Such variability in the operation of the Metonic cycle may give us a reason why Greek civil calendars were notorious for being out of step with each other: it does not mean that they were subject to haphazard intercalation or suppression of months, but rather that they were subject to different systems of intercalation which nevertheless kept the local calendar in step with the moon.[105]

The question remains: to what use was the Antikythera Mechanism put, with all its time-marking facilities? We have already seen that in the early

Hellenistic period, about 300 BC, the festival calendar from Sais in Egypt employed a *parapēgma*, resembling in its details what we know about that of Eudoxos, incorporating a scheme of twelve zodiacal months and then indexed against the native Egyptian calendar. This illustrates better than anything else that has survived the observation by the Roman agricultural writer Columella (*On Agriculture* 9.14.12), that the star calendars (*fasti*) of Meton, Eudoxos and other ancient astronomers were adapted to public sacrifices. Yet the Antikythera Mechanism would be technological overkill for basic agriculture. So what was its purpose?

Astrometeorology is a possible function for the Mechanism, and one with cultural and practical significance. The Hellenistic Tower of the Winds in Athens, now dated like the Antikythera Mechanism to the second century BC, demonstrates beautifully on its exterior walls the combination of solar and meteorological data that was considered useful, while its interior apparently contained an elaborate water clock, whose mechanics still elude us.[106] Nonetheless, despite the monumentalisation of astrometeorology through the Tower of the Winds, the Antikythera Mechanism clearly went beyond such a function. If it was used for this purpose, that must have been just as a means to another end.

Sailing too is a plausible context, if this Mechanism was ever put to use on board ship in the Mediterranean. When we could assume only a single run through the alphabet for the whole zodiac, what little is available of the *parapēgma* on the Mechanism seemed to suit a preoccupation with those times of the year when it was useful to know whether the sailing season could begin (in April–June) and when it should end (in September–October).[107] Now it is difficult to be so sure. Related to this idea is John Seiradakis's recent suggestion that the Mechanism might have been used to help find geographical longitude. While there are several significant assumptions about the meaning of 'longitude', what the *parapēgma* implies by the 'rising' and 'setting' of the stars, and the method used to find 'longitude' from the Mechanism, the idea does show how the different time scheduling dials and inscriptions could be coordinated to produce a result.[108]

I have inclined myself towards an astrological purpose, banal and even objectionable though this may now seem to modern scientists for so complex an instrument. The Mechanism could have permitted the rapid calculation of the positions of all the major planetary bodies, and related phenomena, essential to ancient astrology. Planetary positions are recorded with a remarkable degree of accuracy in surviving tables from the Imperial Roman period, and until now it has not been understood how these positions were so accurately recorded, whether by observation, or calculation, or a mixture of both.[109] The Antikythera Mechanism provides a means. The earliest surviving horoscopes are the sculpted one at Nemrud Dağ in Commagene, Turkey, from 62 BC, and the slightly earlier literary one of 72 BC preserved by the mid-first century AD astrologer Balbillus, who married his daughter into the royal family of Commagene and perhaps thereby acquired access to astrological archives of earlier vintage.[110] Genethialogy, or casting a horoscope for one's birth, must have already developed in the eastern Mediterranean well before the period of the earliest surviving material evidence for it, and it is indeed likely that this type of astrology was being practised in the period now assigned to the Mechanism. Price's argument that omen-reading was the driver for astronomical observation and recording is attractive.[111] If the Mechanism was made for astrological, rather than what we would term astronomical, purposes, it deserves to be read in those terms, rather than to have the astrological function put to one side in favour of an interpretation that simply suits present-day understanding of what 'science' is.[112] Anachronistic triumphalism can undermine our proper understanding of technical innovations in antiquity, where metaphysical and religious concerns were often more powerful drivers. The newly discovered presence of the civil calendar on the Mechanism does not seem to me to speak against an astrological usage, although it does tend to suggest that the instrument was intended for a relatively narrow geographical, or more correctly cultural, region.

There is some evidence that suggests the Antikythera Mechanism was not unique in antiquity. Within the instrument itself it has been thought that there are physical signs that it may be a composite, cannibalising parts of older devices, a circumstance that suggests the existence of other, similarly elaborate, astronomical instruments.[113] Furthermore, we have literary allusions to mechanisms of this or closely related types, which we may loosely call 'planetaria', without necessarily locating the prime purpose of the Mechanism in that category. Outstanding among these devices, which sought to replicate the motions of the celestial bodies, were Arkhimedes' sphere and Poseidonios' orrery, described with such admiration by Cicero (Cicero, *Republic* 1.14.21–22, *Tusculan Disputations* 1. 63, *The Nature of the Gods* 2. 87–8).

Cicero characterises the second globe of Arkhimedes as a technological marvel:

> But this type of globe, on which were set the motions of the sun and moon and of those five stars which are called the planets, or, as it were, the wanderers, could not be represented on that solid globe. And in this the invention of Archimedes was to be admired, because he had thought out how a single revolution should maintain unequal and varied courses in dissimilar motions. When Gallus moved this globe, it happened that the moon followed after the sun by as many revolutions on the bronze, as it does in so many days in the sky. Therefore also that same eclipse of the sun occurred on the globe, and the moon then came to that turning point which is the shadow of the earth, . . .
>
> (Cicero, *Republic* 1.22)

We can certainly sympathise with this level of admiration now, as we see more of the secrets of the Antikythera Mechanism laid bare, and its facility for predicting eclipses demonstrated. But let us not lose sight of the other cause for admiration that Arkhimedes' and other spheres drew from their ancient viewers:

> For when Archimedes fastened on a globe the motions of the moon, the sun and the five planets, he effected the same as that god of Plato, who built the world in the *Timaeus*, so that a single revolution controlled movements dissimilar in slowness and speed. Therefore if in this world things cannot happen without a god, neither could Archimedes have reproduced the same movements upon a globe without divine genius.
>
> (Cicero, *Tusculan Disputations* 1.63)

This time Cicero's admiration is expressed in more philosophical, even theological, terms, with explicit reference to Plato's Demiurge creating the world in his dialogue, *Timaeus*. For Plato, the ordering of the world into regular movements served a deeper purpose, for it was by observing these movements that humans could learn to order and regularise their own interior movements. We are granted our sight, he says,

> . . . so that by observing the circuits of intelligence in heaven, we might make use of them for the revolutions of our own thought, which are related to them, though ours are troubled to theirs untroubled; and that, having learned thoroughly and sharing the ability to calculate rightly according to nature, by imitating the completely fixed revolutions of the god, we might settle the wandering revolutions in ourselves.
>
> (Plato, *Timaeus* 47b6–c4)

On this more philosophical level, Beck has recently pointed out how the far more abstract form of planetarium, which is represented by the Antikythera Mechanism, closely approaches the Platonic ideal. He notes that the instrument's underlying mathematical formula for the Metonic cycle (19 years = 254 sidereal months = 235 synodic months)

> is the intelligible reality behind the relative motions of the visible Sun and the visible Moon . . . But the addition of the little model luminaries is for the purist something of a distraction, a concession to appearances which, even if they can be replicated precisely, are not really worth replicating since in the strictest sense they are unintelligible.[114]

In other words, the Antikythera Mechanism can be seen as a closer approximation of the Platonic, idealised conceptualisation of the cosmos than Arkhimedes' or Poseidonios' orreries. We may add that inasmuch as the Mechanism abstracts the workings of the cosmos to number, so time too is thus abstracted.

That time is number was a tenet of Greek philosophy from Plato and Aristotle onwards. For Aristotle, 'This is time, the number of movement with regard to before and after' (*Physics* 219b1–2). Earlier, in his *Timaeus* (37c6–38c1), Plato had the creator Demiurge fashion the Cosmos in such a way that it resembles, as much as possible, its model, the eternal:

> Now the nature of this being happened to be eternal, and it was not possible to bestow this character completely on the created. He decided to make a moving image of eternity, and when he organised the heaven, while eternity stays in unity, he made this image eternal but moving according to number; and we call this being time. For there were no days and nights and months and years before the heaven came to be, but together with its construction he made the beginnings of them. These are all parts of time, and the 'was' and 'will be' are created forms of time, which we unwittingly but wrongly attribute to the eternal being; for we say that it 'was', it 'is' and it 'will be', but the truth is that only 'is' belongs to it according to the true argument, . . .
>
> (Plato, *Timaeus* 37d3–38a1)

The debate on how to define the eternal being lasts a long time: in the late fourth century AD it is the source of Augustine of Hippo's comment (*Confessions* 11.12.14) that someone once quipped that for those who wondered what God was doing before he made heaven and earth, he was preparing hell. Augustine then provides an extensive disquisition on the nature of time and eternity, reflecting something of the long debate, from Aristotle to the Neoplatonists, about the nature of time. This in itself does not concern us here, since we are focusing on the instruments of time and what they may tell us about perceptions of time.[115]

Let us just note that from a Platonic point of view the sensible world of the visible celestial bodies was a far less accurate depiction of reality than the intelligible world of Forms. Mathematics provided a medium through which one could more closely approach the reality of the Forms. The Antikythera Mechanism, as much as any constructed object could do, brings us that bit closer to metaphysical reality. Cicero would have understood. So, too, would have his contemporary, the astronomer Geminos, who stated:

> It is assumed in all astronomy that the sun, the moon, and the five planets move at uniform speed in circular fashion and in a contrary manner to the cosmos. For the Pythagoreans, who were the first to

> enter into such investigations, assumed the movements of the sun, the moon, and the five planets to be circular and uniform. For they would not accept disorder, with regard to divine and eternal things, such as would make them move at one time more swiftly, at another time more slowly, and at another time stand still (which indeed they call the stationary points of the five planets). For no-one would accept such irregularity of motion even of a decent and orderly man in his journeys. For the necessities of life are often causes of slowness and swiftness for men. But for the incorruptible nature of the stars it is not possible for any cause of swiftness or slowness to be adduced. Therefore, they proposed thus: how the phenomena might be accounted for by means of circular and uniform movements.
>
> (Geminos, *Introduction to Astronomy* 1.19–21)

I think it remains important that we do not lose sight of this fundamental world-view of ancient astronomers when trying to understand the primary function of as complex an instrument as the Antikythera Mechanism. Simplicius, in the mid-sixth century AD, reported that:

> Plato . . . set . . . this problem for all keen about these matters: by what assumed, uniform and ordered movements the phenomena can be saved in relation to the movements of the planets.
>
> (Simplicius, *On Aristotle's On the Heavens* 488.18–24.[116])

From the time of Plato onwards, Greek astronomers were not simply empiricists seeking to describe the apparent phenomena, as we find in the *parapēgmata*, or to predict eclipses and planetary positions, with however complex an instrument, such as the Antikythera Mechanism. Nor did they seek to derive, inductively, a system from those particular, empirical data of star positions, eclipses or planetary positions. Rather, their purpose was to provide a theoretical basis, an overarching system, into which the observable phenomena, especially regarding the planetary system, could be fitted.[117] The Antikythera Mechanism brings us closer than any other surviving instrument from antiquity to a full realisation of this philosophical ideal.

4

TELLING TIME

The earliest, surviving, dedicated sundial technology in the vicinity of the Greek and Roman worlds comes from Egypt and Mesopotamia. In Egypt, from the middle of the second millennium BC, there are L-shaped shadow clocks which measured the shadow cast through the day (though how is still debated).[1] From Mesopotamia, from the end of that millennium about 1000 BC, there are written tables of shadow lengths between sunrise and sunset.[2] These Babylonian tables presuppose some physical means of casting those shadows. In fact, the evidence for sundials among Greece's neighbours is so early that it is surprising that we do not have earlier signs of anything similar from Greece than we do. The absence for so long of such technology in Greece suggests that what the Greeks did develop in this line was at least partly imported, rather than indigenous.[3]

Literary evidence for the earliest stage of the adoption of sundial technology by the Greeks is late, scant and obscure. Diogenes Laertius, writing in the third century AD, mentions the introduction of the sundial to Greece in the form of the *gnōmōn*, a shadow-casting projection of some sort, and associates it with the philosopher Anaximander, who lived in Miletos in the first half of the sixth century BC:[4]

> He also was the first to discover the *gnōmōn*, and he set it on the shadow-catchers (*epi tōn skiothērōn*) in Lakedaimon, as Favorinus recounts in his *Miscellaneous History*, showing the solstices and the equinoxes; he also made *hōroskopeia*.
>
> (Diogenes Laertius (2.1))

Much later still, the *Suda* (s.v. 'Anaximandros') also attributes to Anaximander the introduction of the *gnōmōn* (presumably to the Greeks), as well as the discovery of the equinox, solstices and *hōrologeia*. That a *gnōmōn* at this early stage was a simple, shadow-casting stick stuck vertically in the ground may be inferred not only from later usage of the word, but also from the story that the astronomer, Oinopides, was able to appropriate the term by the late fifth century BC, so as to refer to a perpendicular line as being 'like a *gnōmōn*' when drawn to another straight line from an external point.[5] Diels, followed by Gibbs, thought that Anaximander's sundial took this form so as to provide a measure of only the noon shadow lengths.[6]

This is what a very basic sundial *gnōmōn* would show, if it was used as a calendar for the sun's motion through the seasonal year. But the fact that Diogenes says that Anaximander placed his instrument 'on the shadow-catchers in Lakedaimon' can give the impression that sundials in some other form already existed in Greece, at least in Lakedaimon (Sparta). Just what form these may be thought to have taken is unclear. Informal, indeed mobile, shadow-catching techniques certainly existed in Greece, as we shall see, and they relied on the availability of open, flat spaces, which are not always a ready commodity there. The 'shadow-catcher' (*skiothēras*) is itself specifically identified by Vitruvius (*On Architecture* 1.6.6) as a type of a sundial, comprising an upright *gnōmōn* set up on a flat surface.[7] This flat surface was the shadow-catching area. So perhaps what Diogenes Laertius is telling us is that an open, flat space in Sparta was already set up to 'catch shadows' in some informal way, and what Anaximander did was to set a pole in this ground and thus create a formal, perhaps permanent, *gnōmōn* of fixed height and location – a 'town clock', as it were, which had the capacity to give a standardised measure of time.

We are not told that this *gnōmōn* displayed anything beyond the times of the solstices and equinoxes. Dicks rejected Diogenes' testimony as anachronistic because, he believed, knowledge of the equinoxes required awareness of the sky as a celestial sphere crossed by the main dividing circles of the ecliptic and equator, an idea not hinted at otherwise before the end of the fifth and the beginning of the fourth centuries BC.[8] We can set this criticism aside as misguided. No such conceptualisation of the cosmos is needed to have a rough sense of when or where the equinoxes take place.[9] Establishing their place and time can be, and has been, done easily, albeit roughly, for instance, by marking from a fixed observation spot the halfway point in the journey travelled by the sun along the eastern or western horizon between the two solstices.[10] Assuming, with Diels, that Anaximander's *gnōmōn* was a noon-time shadow-casting instrument, one could then observe the midday shadow cast by it on the four occasions already known from the horizon observations. The solstices would then represent the ends of the shadow-line running north–south along the local meridian, and the equinoxes would be at a point somewhere between them. This gives the simplest meridian 'sundial' (really just a line). Accuracy may be lacking by later standards, but nonetheless the four quarters of the seasonal year can be demarcated without a formal geometric conception of the cosmos.

Quite what Anaximander's *hōroskopeia* ('watchers of *hōrai*') and *hōrologeia* ('readers of *hōrai*') were, and whether they were related to the *gnōmōn*, is hard to tell now. Both Diogenes and the *Suda* separate the references to these instruments from their mention of the *gnōmōn*, as if they really were quite distinct, one from the other. In fact, it is not impossible, given later usage, that they refer not to sundials but to water-clocks, or even to both.[11] The names imply instruments which permitted observation of the *hōrai*, a term that is ambiguous in the context of timekeeping, as it can refer to the annual 'seasons' as well as to the diurnal 'hours'. It would appear to mean 'hours' by the later fourth century BC, when the Cynic philosopher, Diogenes of Sinope, is reported to have been shown an *hōroskopeion* and then commented on how useful it would be in preventing one from being late for dinner (Diogenes Laertius 6.104).[12] That is presumably a comment on the time of day, not on the time of year. But if the *hōrai* implicit within *hōroskopeion* / *hōrologeion* refer to 'hours', then the attribution of an hour-measuring instrument to Anaximander is probably anachronistic by a century and more. If, however, Anaximander's *hōroskopeia* / *hōrologeia* measured only the seasons, then they would seem to be closely related to his *gnōmōn*, which marked the solstices and equinoxes. These four 'tropics' signal the points at which the sun 'turns' (to pick up what the Greek-derived word 'tropic' means), like a charioteer in a race, in its apparent course from one half of the sky, across the celestial equator, on to the other half, and then back again, from one season to the next, in perpetual, annual motion. But the tropical points originally seem to have denoted not the termini of the seasons, as they tend to do nowadays, but the mid-points in their related seasons.[13] Thus, the summer solstice lay at the halfway point in the course of summer, the autumn equinox midway through autumn, and so on. Therefore, while Anaximander's *gnōmōn* may have cast noon-time shadows and showed when the solstices and equinoxes occurred, perhaps his *hōroskopeion* / *hōrologeion* comprised markers on the shadow-receiving surface for the appropriate season which encompassed each tropical point. Later evidence certainly indicates that the term *hōroskopeion* was applied to shadow-casting sundials: not only is this implied by Geminos (2.35, 16.13) and Strabo (2.5.4), but a surviving conical sundial from Pergamon bears an inscription which identifies this particular instrument specifically as a '*hōroskopion*'.[14]

If the story is true that Anaximander invented the shadow-casting *gnōmōn*, and if we understand the testimony correctly that this means a fixed noon-time, calendrical sundial, then this indicates that the technology of the *gnōmōn* appeared in Greece some time early in the sixth century BC. A brief passage in Herodotos, writing a century and a half later in the fifth century BC, credits the Babylonians as the specific source of the *gnōmōn* for the Greeks: 'For the Greeks learned the *polos*, and the *gnōmōn*, and the twelve parts of the day from the Babylonians.' (Herodotos 2.109.3) It is not impossible that Herodotos' 'Greeks' stand for Anaximander.[15] The latter's domicile in western Turkey would have placed him well for receiving scientific knowledge from the east.

All three mechanisms mentioned by Herodotos are, it would appear, to do with sundial technology, if we interpret the terms correctly, though quite how they relate to early sundials is still debated. As we have already seen, the *gnōmōn* may be taken to be a shadow-casting stick, although the word in general signifies basically 'a thing enabling something to be known, observed or verified, a teller or marker'.[16] The twelve-part division of the day looks to be an early reference to the hours, to which we shall return later. Contemporaneously with Herodotos, *polos* occurs in late fifth-century poetry. In Euripides' *Orestes* (l. 1685) and *Chrysippos* (fr. 839.11 N), the word refers to the celestial vault, rather than to the celestial pole or to an instrument.[17] In Aristophanes' *Gerytades* (fr. 163), on the other hand, *polos* is equated with a sundial (*hōrologion*), while in his *Daitaleis* the term refers to the *hēliotropion*, a word of some ambiguity, as other sources suggest that this may be either a form of sundial or a different type of instrument for marking the solstices;[18] either way, it should be a sun-tracking instrument. The *polos* is usually taken nowadays to refer specifically to a spherical sundial, although no ancient source says anything so descriptive about it.[19] The transference of the word from the celestial vault to the spherical sundial would be no great conceptual leap, since that type of dial mirrors the celestial sphere, but all the same this leap still remains hypothetical. Significantly, the assumption that *polos* refers to a sundial is mustered in support of the idea that the earliest type of fixed sundial was hemispherical in form, mirroring the sky. In this view, concave shadow-catching surfaces were an intermediary step along the road towards the development of flat, plane sundials.[20] It is not impossible, however, and perhaps, even more, probable, that concave sundials developed in parallel with and independently of plane dials. This is an issue to which we shall have need to return. Finally, the *polos, gnōmōn* and twelve-part division of the day might be seen as separate entities, or as parts of a single whole.[21] Altogether, they would assist in telling the time both through the year and through the day, providing the basis for both calendars and clocks.

If a *gnōmōn* marked only the solstices and equinoxes, then it marked only the times of the significant changes in daylight in the seasonal year: when daytime is longest, when it is shortest, and when it equals night-time. The *gnōmōn*'s focus is therefore purely calendrical, and as a calendrical device, it is very basic. Whether Anaximander's *gnōmōn* displayed more than the noon shadow in the course of the day, and hence displayed an interest in the sundial as a clock, is not clear from Diogenes Laertius.

But even in marking the solstices and equinoxes, this *gnōmōn* represents only a little advance on Hesiod's division of the sun's year a century earlier. He had noted just the two solstices and neither of the equinoxes in his rough farmer's calendar in his *Works and Days* (479, 564–5, 663). The reason for this lack of interest may lie in what further subdivision of their year the Greeks would have envisaged. They might, like other cultures, have subdivided into two each of the four quarters of the year that the solstices and equinoxes create. There is a hint in Hesiod of such a 'mid-quarter' day when he talks of the sailing season starting 50 days after the summer solstice (663–5), but he does not include all four of these days. By the end of the fifth century, there will be more evidence of subdividing the solar year into eighths through the tropical points and their mid-points, but again the system is not explicit, and we may be expressing things in terms that the Greeks themselves did not recognise.[22]

More to the point is that the Greeks measured the year in their civil calendars not by the sun, but by the moon, and in their agricultural almanacs by the stars. Let us start with the moon. The obvious problem that this generates for sundials is that the moon's cycle is incommensurate with the sun's. In day-lengths, the solar, or seasonal, year is 365¼ days long. No round number of moons can equate with this: twelve moons will fall about eleven days short, while thirteen will overshoot the sun's year by more than 18 days. As we have already seen in chapter 2, the Greeks developed a number of arithmetical schemes to enable their lunar calendars to fall back into line with the sun and the seasons, by adding an extra month every two, three or so years, sometimes in a fixed, repeating pattern. What this means is that a subdivision of the year according to lunar months could not be denoted in a *sun*dial, as an initial year's lunar months would fall out of sync with the sun in the very next year. It takes the development of a notion of a solar month, in the form of zodiacal months, to make it practicable to subdivide the sundial further through the year. That will not take place until the end of the fourth century BC at the earliest.

Before then, when the Greeks did want to mark out smaller divisions of the seasonal year, to remind them, say, when to plough or sow, they did so by observing the risings and settings of the stars. The stars appear to circle around us through the course of the night. From one night to the next, we can observe that they rise or set at gradually different times. So the rising or setting of certain stars can be associated with different seasons or parts of the seasons. In addition, the length of a 'star-year' – the period between, say, a star's first sighting before dawn and the next first sighting before dawn – is very close indeed to a solar year, so close that within a person's lifetime, one would not notice the very slight drift that does occur between them.

Furthermore, the stars have the advantage over the sun as timekeepers that, in any given location, they rise and set over the same place on the eastern or western horizon all the time. They just happen to do this at gradually different times of the day through the year – four minutes earlier each day, in our terms (so for parts of the year they are lost in the daylight). Arcturus, for instance, always rises in the northeast, and sets in the northwest, while Alnilam in the belt of Orion always rises directly in the east and sets directly in the west, and Antares, the prominent red star in Scorpius, rises in the southeast and sets in the southwest.

But, as we saw in chapter 1, the sun rises and sets not only at different times through the year, but also along the same extended arc of the horizon as the three 'fixed' stars just mentioned. The sun is, therefore, initially less useful for timing because it wanders so much in comparison. So it is not surprising to find that our earliest literature in Greece, that of Homer and Hesiod, displays awareness of the utility of the stars as timekeepers more so than of the sun, or even of the moon, both of which they not unnaturally regarded as planets because of their wandering ways. As we have seen in chapter 1 and shall see in chapter 6, tracking the sun at certain times of the year did provide both the Greeks and Romans with a mechanism both for timing particular events and for celebrating cosmological concepts.

To return to the early sundials of Greece: several things probably explain the relatively late entry of fixed dials in Greece as a means of marking the time through the year, in contrast to other astronomical means, such as by the stars or the moon. The stars and the moon provided much sharper focus, although with some inbuilt inconsistencies as far as the moon is concerned. On the other hand, the lack of refined subdivisions in the solar year, such as the twelve months, held up developments of solar timekeepers.

So much, then, for the calendrical aspect of early sundial technology. We are left waiting for developments in the standardisation of technology in the form of sundials, and of time divisions in the form of hours, for any further advance on a simple marker of the tropical points of the solstices and equinoxes. What about the sundial's development as a clock, to mark out time during the day?

As we have already noticed, Herodotos (2.109.3) reports that the Greeks became aware of the *polos*, the *gnōmōn* and the twelve-part division of the day from the Babylonians. This testimony should mean that by his time the Greeks were introduced by their eastern neighbours to the idea of measuring cast shadows to mark at least points in time, if not the passage of time, via a fixed rod, which served as a simple sundial. We have examined this testimony on the assumption that it refers to a noontime dial only. That it could also have referred to time on either side of noon seems a reasonable conclusion considering that Herodotos speaks in the same breath of the introduction to Greece of the twelve-part division of the day. This sounds like an early reference to the twelve hours, for which there does not yet seem to have been a specific Greek word.[23] But Enoch Powell, writing at a time when he was a leading Herodotean scholar, thought that references to sundial technology, in the form of the *polos* and *gnōmōn*, and to the hours, through the twelve-part division of the day, must relate to developments well after Herodotos' time, and so he regarded this passage as a later, Alexandrian interpolation.[24] His argument, however, was awry. The 'hour' as expressed by the word *hōra* first appears in surviving literature in the Hippokratic medical texts of the first half of the fourth century BC, a source untapped by Powell.[25] To the second half of the same century belongs the oldest surviving Greek stone sundial, one unknown in Powell's time: it is an equatorial plane dial engraved with twelve equinoctial hours, from the Amphiareion at Oropos in Attika (Figure 4.1).[26] It is of such sophistication as to imply an earlier

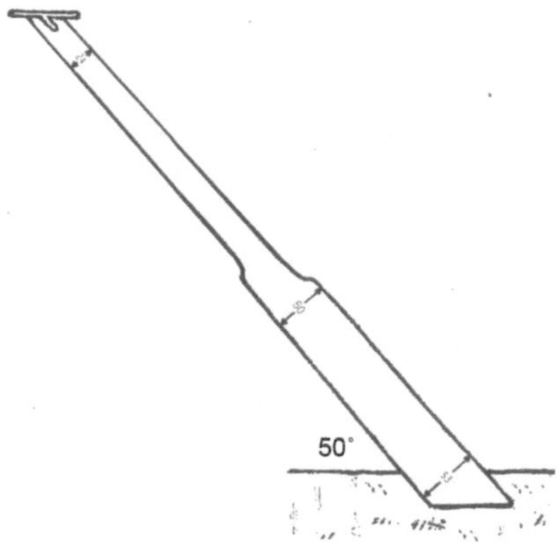

Figure 4.1 Piraeus, Archaeological Museum, inv. 235: plane sundial from Oropos.

Source: Drawing from Schaldach 2006: 112; reproduced by kind permission of Verlag Harri Deutsch.

heritage. So it looks much more likely now that sundial technology and the division of the day into hours could have infiltrated Greek society, at least its 'scientific' community, by Herodotos' time.[27]

But what is meant by 'hour'? In antiquity the hours were a function of dividing daylight or night-time into twelve sections, but there were two means of doing so. One method produced 'seasonal' hours, while the other gave 'equal' or 'equinoctial' hours. Because the length of daylight and night-time in temperate latitudes like Greece varies according to the seasons, dividing the day or night into twelve equal parts produces daylight hours which are necessarily longer in summer than in winter, while the night-time hours are shorter in summer and longer in winter.[28] These hours are called 'seasonal'. A seasonal hour would stretch, in our modern terms, from about three-quarters of an hour in Athens or Rome in midwinter (47 minutes in Athens, 45 minutes in Rome) to about an hour and a quarter in midsummer (1 hour 14 minutes in Athens, 1 hour 15 minutes in Rome). The type of hour that we are used to nowadays, of equal length regardless of the season, derives from the length of the hour specifically at the equinoxes, when daytime and night-time are practically equal. An equinoctial hour may be defined as one-twelfth of the daytime or night-time at the equinoxes, or as one-twenty-fourth of the time from one sunrise to the next at any time of the year. While the earliest occurrences of certainly identifiable seasonal and equinoctial hours in Greek literature belong to the beginning of the third century BC – in the work of the astronomer Timokharis, and in the astronomical papyrus *P. Hibeh 27*[29] – the earliest surviving stone sundial from Greece, that from Oropos, not only has equinoctial hours but also may predate these literary references by as much as fifty years, and indeed implies a prior period of development. But even if we accept that Herodotos' twelve-part division of the day signifies awareness of the hours, whether they were of the unequal 'seasonal' variety, or of the equal 'equinoctial' type, it is impossible to tell.[30]

Gibbs believed that there was nothing to suggest that a *gnōmōn* marking noon, let alone hours, was used by ordinary people, as opposed to the 'scientific community', in Herodotos' time in the late fifth century, and that we have to wait another century for literary and archaeological evidence which shows that sundial technology had been popularly adopted by the Greeks.[31] This is quite true, as far as the archaeological evidence is concerned: the earliest surviving Greek stone sundial with hour lines is currently the plane equatorial dial found in the Amphiareion at Oropos in Attika. It can be dated on epigraphical and historical grounds to the second half of the fourth century BC.[32] Literary evidence furthermore suggests that popular recognition of the hours did not occur until the third century BC, when Kallimachos (fr. 550 Pfeiffer) uses the term *hōra* in this manner to refer to a single hour.[33] Nonetheless, the Oropos sundial is of such a level of sophistication as to imply a period of prior development further back into the earlier fourth century at least. And a familiarity with the underlying principle and potential of the *gnōmōn* for marking out time *within the day* is definitely demonstrated in Greek comedy from early in the fourth century BC. What this potential produces, however, is not 'hours' of time, but instead 'feet'. What remains open to question is at what stage these 'feet' became standardised. Our interpretation of Diogenes Laertius's testimony regarding Anaximander's *gnōmōn* left open the possibility that a standardised form of calendar time could have been available in parts of Greece from the sixth century BC. Standardisation of diurnal time, however, is another issue.

A passage from Aristophanes' *Ekklesiazousai*, produced perhaps in 391 BC, demonstrates the system based on 'feet'. The citizen Blepyros is wondering, in the new, communal world decreed by the women of Athens:

Blepyros: . . .But who will work the land?
Praxagora: The slaves. Your only concern will be to get perfumed and go to dinner, when the *stoicheion* is ten feet.

(Aristophanes, *Ekklesiazousai* 651–2)

The time for a meal is here reckoned from the length of a *stoicheion*. This has been variously taken as another name for an artificial *gnōmōn* or for a person's shadow. Bilfinger long ago demonstrated conclusively, via later scholiast and lexicographic references, that the term must refer to people's shadows, which were measured out literally by foot to give an indication of the time of day. The actual process of stepping out to measure the shadow gives us the derivation for the term *stoicheion*: from *steichō*, 'I walk', in Greek.[34]

How sophisticated was this human sundial in reality? Bilfinger argued that regardless of the height of any person, the ratio of the foot to the height of that individual will remain about 1:6. So it did not matter whether a tall, a medium or a short person measured out the shadow: the resulting number of personal 'feet' would be the same, even though the shadows would be of different lengths according to any standard measure. Let us assume a ratio of 1:6 for a foot to a person's height. This gives us an individual 6 'feet' tall. Let us further assume a 10 'foot' shadow cast by that person. To cast such a shadow, the sun would be at an altitude of $\tan^{-1}[6/10]$, or 31° (ignoring the minor displacement caused by atmospheric refraction) (Figure 4.2).

The sun would achieve an altitude of 31° twice in the day, once in the morning, then again in the afternoon; we need the afternoon reading. We can then calculate, as an example, dinner time for the season when Aristophanes' play was probably performed. If the play was performed at the winter Lenaia festival in about late January, this altitude would correspond to 1:22 p.m. (local solar time), about three hours 40 minutes before sunset. If the performance was around the end of March at the springtime City Dionysia, the great festival in honour of the god Dionysos, when many of the great Athenian tragedies and comedies were first performed, then the sun would be at 31° at 3:33 p.m., about two and a half hours before sunset.[35]

How literally we can take such a source as a comedy is, of course, debatable. And this '10 foot' shadow is based on a particular ratio. Others have argued for a ratio of 1:7.[36] But such ratios are only averages. Real people differ not only in size but in proportion as well. And the equation of such shadows with exact times in modern terms, such as 1:22 p.m., lends an air of precision in timing that is not encountered in popular contexts in the ancient world. Only astronomers display an apparent interest in small

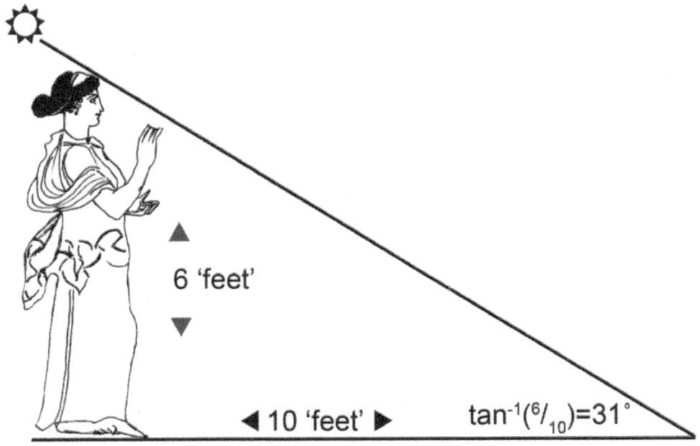

Figure 4.2 A '10 foot' shadow: a person 6 'feet' in height casts a shadow 10 'feet' in length when the sun is at an altitude of 31°.

Source: Drawing by R. Hannah.

subdivisions of the hour.[37] Aristophanes' '10 feet' time would have been meant to provide only a rough estimate of when dinner should be, not a precise measurement.

Even so, fixed shadow lengths throughout the year, such as this '10 feet' time, would lead to a variable time for the same meal. A variable mealtime may well have suited people's biological clocks, since the length of the day itself varied throughout the year. Meals would literally have been movable feasts. The more rural and non-urbanised a society is, the more its rhythms are likely to follow those of nature: hunting, fishing and farming communities must shift their activities to suit the sun and the moon through their seasonal or monthly changes. And we should remember that Athens was little more than a large village in antiquity, so rural rhythms will still have persisted. Aristophanes' plays demonstrate as much for the end of the fifth century BC.

Shadow length as a guide for the timing of activities persisted well beyond the fifth century.[38] According to Athenaios (1.8b–c, 6.243a), the two fourth-century comedy writers, Euboulos and Menander, each had a character who was invited to dinner when his shadow was either 20 feet long (Euboulos) or 12 feet long (Menander), but who arrived too early, because he had been misled by a shadow of similar length cast either by the sun at dawn or by the moon just before it.[39] Let us take the earlier story from Euboulos first:

> Euboulos the comic poet says somewhere:
>
> There are among our guests invited to dinner two invincibles, Philokrates and Philokrates. For even though he is one, I count him as two, great ones . . . three, even! They say he was once invited to dinner by some friend, who told him to come whenever the shadow (*stoicheion*) measures twenty feet, and from dawn he immediately measured as the sun was rising, and when the shadow was greater than by two feet he arrived. Then he said he had come a little earlier because of business, though he came at daybreak!
>
> (Athenaios, 1.8b–c)

At the end of March and with the same parameters as before, a 20-foot shadow on a '6-foot' person would correspond to about 16°40′ for the sun's altitude, which would occur at 4:47 p.m., roughly an hour and a half before sunset. The joke arises, as with Aristophanes, from an unrealistic expectation of how early one should get to dinner, but this time the exaggeration is magnified to the point of including almost the whole day. Philokrates mistakes the equivalent morning shadow, or rather one 2 feet longer still, for the one he should use. A 22-foot shadow corresponds to 7:13 a.m., an hour and twenty minutes after sunrise (exactly 20 feet computes to 7:21 a.m.).[40] So he was ready to dine all day long – hence, presumably, Euboulos' characterisation of him as 'invincible'.

It is possible that between Aristophanes' time and that of Euboulos a later 'hour' for dining had evolved, shifting from mid-afternoon to early evening. The slightly later testimony from Menander further complicates this issue of when dinner time was, not only because he uses a different measure of 12 'feet', but also because he uses the moon, and without indicating which phase. A brief investigation is in order:

> And in the *Temperament*, he [Menander] says:
>
> The man, whoever he is, differs not a bit from Chairephon, who, invited once to a dinner at twelve-foot, at dawn seeing the shadow from the moon ran as if he was too late, and arrived at daybreak.
>
> (Athenaios, 6.243)

Again, we do not know the exact parameters for the time of year, and that matters more in this instance because of the variability in the moon's position. But we can impose some limitations by having the moon waning a few days after being full, since at full it sets as the sun rises and whatever light it casts is swamped by the sun's at that moment; and because when it is still waxing before full, the moon achieves the altitude appropriate for the story nearer midnight than dawn, too long before sunrise for the joke to be sensible. So, to take the spring City Dionysia around late March as an example again, we need the waning moon to be at an altitude of about 26°30′ to create a shadow of 12 feet from a '6-foot' individual. Let us take the day when the moon is 18 days old, four days beyond full. The moon would be high enough to cast a 12-foot shadow in the dark at about 3:55 a.m., two hours before sunrise. For comparison, the expected dinner time, when the sun would be creating the 12-foot shadow, would be about 3:58 p.m., two and a quarter hours before sunset.[41] Once again, we have an accidental, or rather avaricious, diner, who arrives far too early for the appointed meal presumably in the hope of all-day dining.

But some have found this idea of a variable mealtime hard to swallow and have assumed instead that mealtimes would preferably have been at a fixed time of day. Edwards thought that the variability in timing through the year would be such that a remedy must have been sought. This remedy, he suggested, was found initially by adopting varying shadow lengths for different times of the year and then ultimately by using 'seasonally determined shadow tables'.[42] These would obviate the need to consult one's shadow at all, or to vary one's meal time much. But problems arise with this hypothesis. The idea that shadows of different length were adopted for different periods of the year is based on the varying lengths mentioned in the literary sources that we have already examined. Yet it is possible that these variations in length are not seasonal, but instead serve different purposes, as jokes raising laughter from different forms of exaggeration.

Furthermore, while there is certainly evidence for Greek shadow tables, it is very late and their linkage back to the pre-Hellenistic Greek world is tenuous. Of the surviving Greek tables, about twenty altogether, only two are antique, the earlier dating to ca. 200 BC, the other of Roman date; the rest are Byzantine from the twelfth to the sixteenth century. Other tables survive from Late Antiquity and the medieval period in Latin, Coptic, Ethiopic, Arabic, Syriac and Armenian, but derive, Neugebauer believed, from the same Greek prototype as the Byzantine examples.[43] The tables fall into two main categories: those which organise the seasonal shadow lengths according to the zodiacal months, and those which group them by calendar month. The surviving forms would make the tables Hellenistic in origin at the earliest, since the zodiacal month is an invention after ca. 300 BC, and the calendar months – Julian or Alexandrian – are later still.[44]

One of the Byzantine tables gives a flavour of the type of entries that occur in the tables. The text is presented as a letter, dedicated by one Sextus, the *hōrokrator* ('ruler of the hours'), to a King Philip. In the introduction Sextus tells the king:

> Whenever you want to know whatever hour it is, I will tell you here: in the place where you may be walking about, you must measure out your own shadow yourself; and when you have found the shadow of your head, mark the place and begin to walk from where you are standing, one foot in front of the other; and see also how many feet you make, look in the work of the walls(?), on which the months are inscribed; and so you will find in each month the hour of the day.
>
> (*CCAG* 7.188.4–189.2)

The text continues with first a list of the characteristic differences between each hour in feet, and then a table of each zodiacal month's hours in feet. So we are told that:

> whenever it is the first hour, always [reckon] that it is ten feet longer than the second; the second is four feet longer than the third hour; the third three feet longer than the fourth hour; the fourth two feet longer than the fifth hour; the fifth one foot longer than the sixth hour; and the sixth hour is the base-line.
>
> (*CCAG* 7.189.2–6)

The sequence from the sixth hour to the twelfth then uses the previous differentials, but of course in reverse order, as the shadows lengthen to evening. There follows a table of the zodiacal months, arranged in pairs which bear the same shadow lengths: Sagittarius and Capricorn; Scorpio and Aquarius; Libra and Pisces; Virgo and Aries; Leo and Taurus; Cancer and Gemini, e.g.

When the Sun is in Sagittarius and in Capricorn

hour 1	feet 28	hour 2	feet 18
hour 3	feet 14	hour 4	feet 11
hour 5	feet 9	hour 6	feet 8
hour 7	feet 9	hour 8	feet 11
hour 9	feet 14	hour 10	feet 18
hour 11	feet 28	hour 12	feet 0

When the Sun is in Scorpio and in Aquarius

hour 1	feet 27	hour 2	feet 17
hour 3	feet 13	hour 4	feet 10
hour 5	feet 8	hour 6	feet 7
hour 7	feet 8	hour 8	feet 10
hour 9	feet 13	hour 10	feet 17
hour 11	feet 27	hour 12	feet 0

(*CCAG* 7.189.11–24)

The variation of foot-length, and therefore of hour-length, from one zodiacal month to another indicates that these hours are seasonal, not equinoctial. Other tables in the calendar-month mode, however, use equinoctial hours.[45] So it seems that the tables evolved over time and region.

Neugebauer saw in 'King Philip', the dedicatee of this and other tables, signs of a medieval mangling of a Classical astronomer's identity, namely the Philip referred to in some *parapēgmata*, alongside Euktemon.[46] For this reason, he thought the origins of Greek shadow tables lay in the fifth or fourth century BC. As we shall soon see, though, this is very unlikely, and a later date in the Hellenistic period looks more plausible.

The method in the tables, of pacing out one's own shadow, undoubtedly fleshes out what we would guess lies behind the references in earlier Greek comedies. But the difference with the tables is that the shadows are then organised according to arguments – zodiacal or solar calendar months – which did not exist in Classical Greece. While shadow tables of much older vintage existed in Egypt, the earliest from the Middle Kingdom,[47] and could have been an influence later in the expanded, Hellenistic Greek world, the existence of a solar calendar in Egypt, with its twelve 30-day months (plus the five 'epagomenal' days tacked on at the end of the year) assisted greatly in the organisation of the seasonal shadows into evenly sized and consistently regular blocks of time. A pre-Hellenistic prototype in the Greek world would inevitably be hampered by the lack of any such organising argument. As we saw with the early *parapēgmata* in Greece, before the introduction of zodiacal months the most likely organising principle was the day-count, from one phenomenon to the next; but even that option is controversial. Such a system would be a very irregular argument against which to organise shadow tables: we need the concept of evenly sized packets of time to take hold in Greece before a shadow table becomes a reasonable and practicable option. Until then, it looks more likely that shadows of fixed length held throughout the year, causing events timed against them to wander up and down the day through the seasons. Furthermore, for this reason it is improbable that Greek shadow tables, as we have them, could stem from the fifth or fourth century BC, when not even zodiacal months had yet been invented. So Edwards' proposed evolution, from fixed shadow lengths throughout the year, through different shadow lengths for different times of the year, to seasonal shadow tables, fails to pass scrutiny in detail.

Nevertheless, some sort of development from fixed shadows to seasonally varying shadows captured in tabular form remains necessary at some later stage. Why would this have happened? If the adaptation of shadow lengths to tables organised around zodiacal months took place some time from the late fourth century BC onwards, this development occurred in a period when fixed stone sundials were also being made. Did plane sundials, or more specifically simple meridian lines, trigger the development of the shadow tables? A small meridian line from Chios, for example, illustrates the use of monthly zodiacal divisions along the noontime line in the second century BC.[48] But problems arise: the shadow tables reflect the persistent use of the human body as a *gnōmōn*, rather than of a standardised, artificial rod; and they present shadow lengths for more than just noontime. One might surmise that after the creation of zodiacal months at some stage in the Hellenistic period, there followed a growth of interest in marking out stone dials with monthly divisions. Yet this is rare among surviving dials, which demonstrate a greater interest in the subdivision of the day into twelve hours, rather than of the year into twelve months. Nonetheless, meridian lines (if not plane sundials) do show an interest in indicating the twelve months – witness the Chios line in the middle of the Hellenistic period, and the huge meridian line of Augustus in Rome at the end of it.[49] Is it a coincidence that the shadow tables resemble such plane-surface technology, as opposed to systems based on the sphere? Schaldach suggests that we should no longer see sundials developing from the spherical as the prime ancestor – a notion based on a fixed idea that the *polos* was a hemispherical sundial – but from plane dials, of a type initially introduced from Egypt or Babylonia.[50] The conclusion seems inescapable as far as plane sundials are concerned; spherical sundials, on the other hand, seem to demand a different origin, derived from considerable reflection on the apparent hemispherical form of the sky.

We saw with Anaximander that the possibility existed for a standardised time to be read off a fixed *gnōmōn* from as early as the sixth century BC. But nothing seems to have come of it in the public sphere. Seasonally influenced timing of other activities remained characteristic of Greek society, from the agricultural to the political – the Assembly in Athens, for example, began at daybreak, announced by cock-crow (Aristophanes, *Acharnians* 19–20, *Ekklesiazousai* 30–31, 82–85, 289–92, 390–91), a shifting timetable if ever there was one. The very notion of a fixed time for meals is also pilloried by satirists as being the direct result of the introduction of slave-driving sundials. Aulus Gellius, in the second century AD, relays the sentiment via a quotation from Plautus, three hundred years earlier:

> The gods damn that man who first discovered the hours, and – yes – who first set up a sundial here, who's smashed the day into bits for poor me! You know, when I was a boy, my stomach was the only sundial, by far the best and truest compared to all of these. It used to warn me to eat, wherever – except when there was nothing. But now what there is, isn't eaten unless the sun says so. In fact town's so stuffed with sundials that most people crawl along, shrivelled up with hunger.
>
> (Aulus Gellius 3.3.5, attributed to Plautus)

The comic *topos* resurfaces a little later in the fictional letters of dinner-seeking parasites, written by Alkiphron around 200 AD:

> The *gnōmōn* isn't marking the sixth hour yet by its shadow, and I am in danger of withering away, goaded by hunger. So, Lopadekthambos, it's the hour for you to have a plan, or better still a crowbar and a rope to get hanged with. For if we bring down the whole column that holds up this hateful sundial, or turn the *gnōmōn* this way to bend where it will be able to indicate the hours sooner, that will be a plan worthy of Palamedes! Now, I'm telling you, I'm dry and parched from hunger. And Theokhares doesn't take to his bed until his slave runs to tell him that it's the sixth hour. So we need some sort of scheme that will be able to outwit him and upset Theokhares' habits. Brought up as he has been by a stern and frowning pedagogue, he has no new ideas, but like a Lakhes or an Apolexis he is strict in his ways and doesn't let his stomach get filled before the hour has come.
>
> (Alkiphron, *Letter* 3.1)

Both Alkiphron and Plautus tended to reuse material from Greek New Comedy of the later fourth century BC, particularly from Menander. It is in a fragment from Menander that we first hear of the half-hour, an unprecedented level of precision in ancient timing among the general public. We cannot know the literary context of that fragment now, but we may be able to recover the scientific context in which the half-hour developed. As we have already seen, the Cynic Diogenes of Sinope, who lived in the second half of the fourth century BC, is reported to have been shown a sundial (*hōroskopeion*) and to have discerned its value as a timekeeper for meals (Diogenes Laertius 6.104). The nature of the story suggests that such sundials were a novelty, something still to be marvelled at. Vitruvius' list of inventors of different types of sundials (*On Architecture* 9.8.1) includes some who lived around this same period: certainly Berosos 'the Chaldaean' and possibly Patrokles.[51] So the anti-sundial joke probably stems from the very period when stone sundials started to become popular in Greece, and with them the idea of the hour and the half-hour. To this same time, at the earliest, we may assign the development of shadow tables, organised according to the newly devised zodiacal months. We may further guess that by about 300 BC, in Menander's time, the informal measure of time via one's own shadow was competing with the measure provided by these tables and by fixed sundials. It would then seem to be from the late fourth century BC or early third century BC that fixed times for activities developed, to the chagrin of some people.

We may wonder whether the increasing refinement of sundials as clocks rather than as just calendars drove the changes in people's habits, or whether the reverse was true, that there were some other social pressures which drove the development of timekeeping technology. If later periods are any gauge, the expectation would be that the impetus came from social changes, not from the technology itself. In Europe from the fourteenth century the growth of the cities and hence of their administrations created an increasing need to timetable meetings and other activities more carefully than had been the case in the earlier medieval world. So meeting times were regularised to certain days of the week and certain hours within those days, a systematisation that was borrowed from the legal world, which itself borrowed its segmentation of the day from the world of the church, with the hours of Terce, None and Vespers figuring prominently. Clock-time appears in documents from the fourteenth century, and increases through the fifteenth and onwards.[52] With this development of clock-time there develops also an increasing abstraction from nature-based means of marking time and hence a separation of previously related activities, such as secular and sacred meeting times, i.e. when one could go to civil meetings as opposed to when one should go to church. Clocks, as abstractions of the celestial order themselves, aided and abetted these processes of abstraction and separation, but did not cause them. Similarly, in the Greek and Roman worlds we should look to social change as the driver for the establishment of fixed times for activities. In the late Classical and early Hellenistic periods, for instance, when we see the first signs of such change along with the technological means to impose it through sundials and water clocks, we might posit the growth of city leagues before Alexander's conquests and certainly of extensive empires after them as catalysts for the development of mechanisms to time activities more precisely and in a coordinated fashion.

The needs of the human body – for food, relief, sleep – provide the simplest marker of the passage of time. This is as true today as it was in the ancient world, however much we allow technology, business and artificial timetables to order our days and so to control these basic demands. The moment it was decided that a fixed hour should serve for mealtimes, or for any other activity, a tension was created between people's physical or social needs and mechanical demands. The story of how the Greeks and Romans learned to measure time is a history of the gradual distancing of people from nature, of their 'denaturisation', as Turner put it.[53] With fourth-century Greek comedy we obtain a glimpse of the beginnings of this process, and with second-century BC Roman comedy we witness something of the upset it created. The reprise of the sentiment in the second century AD may suggest that it was still possible to generate a degree of nostalgia over the loss of natural, bodily timekeepers. But the process of denaturisation seems complete by the sixth century AD, when Cassiodorus points out to the barbarian Gundobad, king of the Burgundians, that the ability of humans to reckon time through sundials marks them out from the animals, which are ruled by hunger:[54]

> Have in your country what you have seen once in Rome. It is proper that your grace should enjoy our gifts to the full, because you are connected with us by affinity. It is said that under you Burgundia examines the most subtle things, and praises the inventions of the ancients. Through you she lays aside her heathen nature, and while she respects the prudence of her king, she rightly desires to possess the inventions of the wise. Let her distinguish the periods of the day by its own movements, let her organise the moments of the hours most appropriately. The order of life passes in confusion if such true discrimination is unknown. Certainly it is the life of beasts to know the hours from the stomach's hunger, and to have no certainty, which is meant to be joined to human customs.
>
> (Cassiodorus, *Letters* 1.46.2)

Nonetheless, even in the organised monastic world inhabited by Cassiodorus and regulated by him and his sixth-century contemporaries, such as the anonymous author of the *Rule of the Master*, Benedict of Nursia and Gregory of Tours, nature still obtrudes in the use of star observations at night and of the cock-crow to signal the approaching dawn.[55] 'Denaturisation' is a relative term. In the ancient world, urbanisation took a long time to come to mean also 'deruralisation', or the removal of the countryside from the town. In the first century AD, whether in provincial Jerusalem or metropolitan Rome, the cock-crow still warned of daybreak (e.g. Mark 14.30, 72; cf. Martial, *Epigrams* 9.68), as it had done for the Athenians in the fifth century BC (Aristophanes, *Ekklesiazousai*, 30–1, 390–1). Even in Rome, when Martial (*Epigrams* 12.57) complained of the emperor Nero bringing the countryside into the city (*rus in urbe*) with his Golden House complex, it was still much less than a day's walk from the city centre to the farmed fields outside the walls.[56] Ancient cities remained within sight and sound of the countryside, and of its timekeeping practices.

If we turn our attention back to the actual sundials, and if we are dealing with a fixed *gnōmōn*, rather than a mobile individual, casting shadows whose length signals the time for activities, what sort of 'clock face' would this create? If such shadows were mapped out on the ground at different lengths, they would trace out arcs of concentric circles. While a disk, inscribed with concentric semicircles signalling the timing for various events, might seem a plausible form for the original horizontal sundial face, no such example has survived to substantiate the hypothetical type.[57] Protractor-like dial faces do survive, but they are late (from the first centuries of our era to well into the western European and Byzantine medieval period), they do not have concentric circular arcs, and they were intended to be mounted vertically rather than horizontally. They are usually regarded as being of limited or no use as sundials,[58] but this is partly because what functionality they offered has been misunderstood, and partly perhaps because of an unscientific interest which they reflect in an ideal time which was not really of this world.[59] In fact, they can demonstrate a very similar approach to time as the human shadow 'clocks' of Classical comedy, and one may therefore wonder why earlier examples have not survived.

As we have just seen, a fixed shadow length – 10 feet, or 12, or whatever – signalled the appropriate time for a certain activity: dining, in the cases we have been examining, but potentially anything regularly undertaken. When seen in terms of our modern, equal hours, the time for that activity necessarily shifted according to the season, falling earlier in the day in winter, but later in summer, as the day itself shortened or lengthened from one seasonal extreme to the other. Dinner time was not at a fixed hour (such as the 'eleventh' hour of the day), because the shadow length would then have varied with the season; but instead dining took place at a fixed shadow length, which translates to a different hour of the day (in both ancient 'seasonal' hours and in modern equal hours). The shadow, therefore, is an 'event marker', to borrow a term, in the same way that the horizon suntracking method was. The protractor-like sundials from antiquity and the 'scratch dials' on churches of the medieval period are in fact 'event markers' too, rather than proper sundials.[60]

These sundials take the form of semicircles (sometimes extended uselessly in the Western medieval examples to full circles, sometimes in the Byzantine examples elaborated artistically), inscribed on flat slabs or walls, and were intended to be set vertically facing south (more or less) (Figure 4.3).[61] The *gnōmōn* was fixed horizontally from the centre of the circle, so as to cast a shadow which moved through the day across equally divided sections of the semicircle. The dials are often regarded as defective, either because (in the elaborate Byzantine examples) by making the sectors in the functional semicircle all equiangular ($12 \times 15°$) they oversimplify the length of the seasonal hours through the year; or because (in the Western 'scratch dials') by carelessly and irregularly spacing out the hour lines, they distort the passage

Figure 4.3 Istanbul, Archaeological Museum, inv. no. 905T: Byzantine vertical plane sundial, 9th–10th centuries AD.

Source: Photograph R. Hannah.

of time according to any regular standard. But as Mills has pointed out, if everyone in a group consults such a dial with a view to taking part in an activity when the shadow falls in a given sector of the circle, then it still allows the group's members to coordinate their activities, at a certain 'hour'.[62] In terms of a standardised measure of time, this activity will take place at different moments of the day as the seasons progress. But that may not be a problem to the group. Dining would have been a movable feast in antiquity, following the vagaries of the seasons through the year. So too would have been the time of prayer in the medieval Christian world, if it followed the 'time signals' of the church 'scratch dials'. It is rather like people nowadays synchronising their watches but to the same wrong time: the 'time' may be inaccurate according to an objective standard, yet it will be 'correct' according to an internal agreement and therefore perfectly adequate for internally coordinating an activity.

At some stage, a type of sundial was developed that showed the passage of the sun, via the shadow of the *gnōmōn*, into each of the twelve zodiacal signs in the course of a year. It is easy to imagine how helpful such a basic type of sundial would have been in marking out the passage of the seasons through the year, and even of dividing the year into smaller, manageable chunks in the form of the twelve zodiacal or solar months. By the fifth century BC a set of twelve constellations had been marked out along the path that the sun appears to traverse in the course of a year: Aries, Taurus, Gemini, and so on, as they are known today under the Roman forms of their names, though the system emanated from Babylonia.[63]

A century ago there developed a belief that the type of sundial called the *arachnē* by Vitruvius (*On Architecture* 9.8.1) and attributed by him to Eudoxos as the inventor was a sundial which showed the passage of the *gnōmōn*'s shadow into each of the zodiacal signs. The assumption was that the *arachnē* must have displayed the full array of criss-crossing lines – solstitial, equinoctial, zodiacal, and hourly – because *arachnē* is Greek for a spider's web, and so the name could easily refer to such a network of time lines on stone sundials. The attribution to Eudoxos would place the invention in the first half of the fourth century BC.[64] But how many lines make a spider's web? In particular, is there any need for the name *arachnē* to imply the addition of the zodiacal divisions in addition to the solstitial and equinoctial lines? On the basis of what has survived, the answer is probably none. The earliest surviving Greek sundial, which also happens to display a simple, weblike network of lines, is the equatorial dial from Oropos. It is dated to the second half of the fourth century BC, just a few years after Eudoxos' lifetime, and not surprisingly it has recently been thought also to illustrate the *arachnē* sundial.[65] It presents the hour lines, but only one solstitial line and the equinoctial lines. There is no indication, even along the meridian alone, of a twelve-part division which would correspond to the zodiacal months. In fact, the link between Eudoxos and a sundial which used the zodiacal signs looks untenable. For calendrical and astrological purposes the full circuit of the zodiac was divided up into twelve equal divisions of 30° each, which were named after their resident constellation. But the distinction between actual zodiacal constellations of varying size and artificial zodiacal signs of an even 30° of arc is not attested in extant Greek texts before the third century BC.[66] It is therefore probably only from about 300 BC at the earliest that we could expect evidence of zodiacally structured sundials. The *arachnē* sundial probably signifies simply the introduction of a dial face inscribed with at the most the three seasonal lines (two solstitial and a single equinoctial) crossed by the hour lines, much like the Oropos dial.[67]

A spherical sundial found in Ai Khanoum in Afghanistan, and datable to the third century BC or first half of the second century BC before the city's sacking, is inscribed with a network of lines indicating the daily hours and the zodiacal sign boundaries (three of which do double service for the solstitial and equinoctial lines).[68] But, depending on how one imagines sundials evolved, it is possible that the first dials to show the sun moving through the zodiac were much simpler than this and similar to the meridian line found on Chios and dating perhaps to the second century BC.[69] This is just a small, flat north–south, noontime line, with the summer and winter solstices marked at the extremities, together with the divisions between the twelve zodiacal signs. The *gnōmōn*'s shadow marked the passage of the midday sun through these solar months. This line represents, in fact, just the type of development that we anticipated from Anaximander's rudimentary *gnōmōn*, if the meridian line was to be more useful as a calendar through the year. So how useful is this one? The *gnōmōn* itself would appear to have been 72 cm in height, or just under half the height of an ancient Greek person. The change in shadow length from midday 24 March (the time of the spring equinox and of the entry of the sun into the zodiacal sign of Aries at the time of the dial's construction) to midday 24 April (the time of its entry into Taurus) would theoretically be just under 21 cm, which is an easily noticed movement.[70] Naturally, if we were using a *gnōmōn* of any greater height, the resultant shadow would be correspondingly longer, as would the meridian line be longer.

But how useful would the Chian *gnōmōn* be from one day to the next? Unfortunately, at the latitude of Chios (38°22′N), the shadow measured by a *gnōmōn* of 72 cm changes too little to help much in the distinction of one day from the next. Between midday 23 March and midday 24 March, for example, the difference in shadow length would be just 8 mm, less than the width of an adult finger.[71] Even with a human-sized *gnōmōn* of, say, 1.50 m height, the change is only 1.6 cm.[72] At other times of the year outside the equinoxes, the movement of the shadow would be even smaller. Given this often imperceptible daily change, a meridian line displaying the shift of the noontime shadow would be useful on a daily basis only if it was much taller than a human figure. Such a sundial, then, is of calendrical use over monthly rather than daily periods.

By the second century BC the Chian meridian line and the Ai Khanoum spherical dial were just two of a number of more or less complex types available to the Greeks and Romans. Vitruvius himself (*On Architecture* 9.8.1) lists fourteen types, along with their supposed inventors, the earliest Eudoxos. The principal kinds surviving in the archaeological record are the spherical, cylindrical, conical and plane. There are other types known, including eventually small, portable ones, which were functional over much of the Roman Empire – a remarkable feat, considering the dials were latitude-sensitive.

The spherical type of sundial was the most labour-intensive and difficult to construct, as it entailed carving out initially a hemisphere but usually a quarter-sphere of stone. It was, however, theoretically the simplest to mark out, because it captured the celestial dome inversely on a matching concave surface (although in practice carving regular curves on the interior surface would never have been easy). Its *gnōmōn* hung out over the hollow part-sphere. The earliest surviving spherical dial is a fragmentary example from the Greek colony of Istros on the Black Sea coast in Romania, which has been dated to the third century BC on epigraphical grounds.[73]

Although not one of the largest spherical sundials, an example in Selçuk in Turkey but said to be from Aphrodisias, is still imposing (Figure 4.4).[74] The

Figure 4.4 Selçuk, Archaeological Museum, inv. no. 375: spherical sundial perhaps from Aphrodisias.

Source: Photograph R. Hannah.

dial alone is 58 cm in height and 1.18 m in width; with the preserved base, the ensemble stands about 93 cm high. Its present *gnōmōn* is a modern restoration which has, unfortunately, destroyed any trace of the original hole for the rod. The length of the *gnōmōn* might be reconstructed, if we knew the precise latitude for which the sundial was originally made and could identify the equinoctial line with certainty – and if the dial was constructed accurately. Unfortunately, none of the above applies in this case.

All in all, if the dial were stood facing directly south in its present setting, we still could not trust its time. As it is, the sundial has been stood facing another direction, not directly south. Nonetheless, several features of this dial are worth noting. Typically, such sundials will present two lines marking out the extreme limits of the lines tracked by the shadow of the *gnōmōn*. These are the shadows of the winter and summer solstices. An intermediate line, which does service for both equinoxes, often lies in between. Other lines may be added for other days in the year, sometimes for the change of months, sometimes for particular festivals.[75] On the Selçuk dial we find certainly two clearly cut lines near the outer edge, while traces of possibly a third line can be made out closer to the *gnōmōn*. The outermost line would usually represent the time of the summer solstice, the innermost that of the winter solstice, and the intermediate line the equinoxes. But as Gibbs has demonstrated, these lines do not correspond exactly with where they should be by calculation. A lack of care about the precise positions of these lines is not unusual.

On the Selçuk sundial eleven hour lines have been marked, criss-crossing with the solstitial and equinoctial lines. In the spaces between the hour lines are the twelve seasonal hours of the day, numbered in Greek fashion: Α Β Γ Δ Ε S Ζ Η Θ Ι ΙΑ ΙΒ. Since the lengths of the hours vary according to the season, the hours on the dial face increase in width the closer they are to summer at the bottom of the dial face, and decrease as they approach winter at the top. If we imagine the sundial facing directly into the sun to the south, with observers looking north into the dial and with the sun at their back, the sun would then rise in the east on their right, casting the first hour's shadow from the *gnōmōn* to their left (A). It will cross the sky behind them from right to left, passing south at midday (the line between S and Z). It will finally set in the west on their left, casting the twelfth hour's (IB) shadow to their right. One final feature of the sundial in Selçuk is the dedicatory inscription on the base of the dial. It tells us that the dial was dedicated to:

Emperor
Caesar M. Aur[elius]
Severus
Antoninus
Aug[ustus] and Julia
Augusta (his) mother . . .

i.e. to the Emperor Caracalla and his mother Julia Domna, thus giving us the general date of AD 198–217 for this timepiece.

It is generally assumed that Vitruvius (*On Architecture* 9.8.1) is referring to this spherical type of sundial when he writes about the *scaphē* or *hemisphaerium*: the latter name particularly recommends the identification. He ascribes its invention to Aristarkhos of Samos, an attribution which would place the invention in the early third century BC, since Aristarkhos (or his 'school') is associated with a summer solstice observation in 280 BC (Ptolemy, *Almagest* 3.1 (H206).[76] This is too late a date for those who assume that the *hemisphaerium* is identical with the *polos* mentioned as early as Herodotos (2.109.3).[77] Soubiran, while allowing for a simple error of attribution on Vitruvius' part, also suggests that while the type existed long before Aristarkhos' time, he may have derived the theory that underlies the type, or (following Ardaillon) he may have transformed the type from a stone model to one of metal.[78] Edwards, on the other hand, allows for the possibility that the type may have enjoyed several introductions to the Greeks. One, he believes, may have come via Egypt, and is indicated in *P. Hibeh 27*, where the author tells of being taught by a wise man of Sais who used the '*holmos ho lithinos* [the stone hollow?] which the Greeks call a *gnōmōn*'.[79]

The conical type of sundial is a variation on the spherical, representing a simpler, partial hollowing out of the stone block. The type is well illustrated by an elaborate example from Pergamon in Turkey (Figure 4.5).[80] The formal similarity to the spherical type is clear. Here the dial has been provided with an artistic support in the form probably of the thematically relevant figure of Atlas, the Titan condemned perpetually to uphold the heavens.

In the surviving archaeological record, the conical competes with the spherical as the most popular type.[81] The earliest surviving example is from Heraclea ad Latmum in Turkey, and dates probably to the end of the third century BC.[82] This might be very close to the period of the type's invention. As we have already seen, at the start of his discussion of sundials Vitruvius (*On Architecture* 9.8.1) attributes the invention of the *arachnē* sundial to Eudoxos, the fourth-century BC astronomer. We also saw that this type of

Figure 4.5 Bergama, Archaeological Museum: conical sundial supported by Atlas(?), from Pergamon.

Source: Photograph R. Hannah.

dial may have been a plane dial, whose receiving face was crossed by no more lines than the three seasonal ones for the solstices and equinoxes, plus the intersecting daily hour lines. But Vitruvius offers an alternative attribution for the *arachnē*, which would associate it instead with the conical type: this is to Apollonios, who is presumably the mathematician of ca. 200 BC, and who wrote on conic sections. This work on the theory of conic sections suggests that Apollonios may have had something to do with the exploration of sundials, given that one of the main types is conic, and that from them he may have derived the theorems for conics.[83] The popularity of the conical type may occasion some surprise, given the apparent complexity of its theory and the greater difficulties in marking out the requisite interior lines, as these are now projected on to an awkwardly shaped curved surface – theoretically, for instance, the hour lines should be double-curved and sinuous. But it is likely that the conical was much easier to construct in stone, because its generating line was straight, not curved as for a spherical dial, while its theoretical underpinning was kept to a minimum, and indeed obviously simplified or even not understood by many of the makers, to judge by their inaccuracies.[84]

The plane type of sundial, which occurs usually in horizontal or vertical forms but also uniquely in a slanted, equatorial form at Oropos, is the easiest to construct but technically the most difficult to mark out. The difficulty arises from the projection of the hemispherical dome of the sky on to a completely flat surface. A shadow, which tracks the movement of the sun through the year, is cast by a *gnōmōn*, which is usually stuck perpendicularly into the flat surface of the sundial (but at an angle in the Oropos specimen, because of the initial slant of the dial face). For Gibbs, the earliest preserved planar dial was 'probably' the horizontal, rectangular, marble slab, which was discovered in a private house on Delos and which dates to the second century BC.[85] But the vertical plane dials on the eight sides of the Tower of the Winds in Athens, built by Andronikos from Kyrrha in Macedonia, are now dated to the second century BC (Figure 4.6).[86] They are therefore among the earliest surviving examples of the vertical plane type, but their complexity and accuracy suggest an older ancestry.[87]

The Oropos sundial, dated to the second half of the fourth century BC, now confirms this earlier heritage back to the fourth century BC.[88] On the basis that the *pelecinum* type of sundial, as its name partly suggests, was a 'double-axe' version of the horizontal plane sundial type, and on the assumption that the Patrokles to whom the invention of the *pelecinum* is attributed by Vitruvius (*On Architecture* 9.8.1) was the geographer associated with the early Seleucid rulers, some have argued for a possible date of invention of the plane type in general between the second half of the fourth century and the early third century BC.[89] Cam, however, disputes the identification of the *pelecinum* with the horizontal dial type, and instead, on the basis of the discussion of it by Cetius Faventinus (29.2), she associates it with a 'two-leafed', vertical type of sundial well-known from later art.[90] If, on the other hand, the

Figure 4.6 Athens, next to the Roman Agora: plane sundial on the Tower of the Winds, southeast side.

Source: Photograph R. Hannah.

geographer's early date had put some people off the attribution of the *pelecinum* to the fourth century BC,[91] the Oropos equatorial dial now warns us that we can still be surprised by new archaeological discoveries into realising that the development of some sundial types really did take place much earlier than used to be thought. So a fourth century BC date for the invention of the *pelecinum* – whatever form it took – is not impossible. And a similar date for the invention of the plane dial – even if it is not the *pelecinum* – is now necessary.

Even among miniature portable sundials, there is complexity. Some of these dials are simply very small versions in limestone of the stone spherical or conical types.[92] Even on one of the smallest and earliest of such miniatures, an ivory conical dial only 2.8 cm in height dating probably to the first century BC, the accuracy of the hour lines is remarkable: it was made for a latitude of 33°, which corresponds reasonably well with the latitude of its provenance, Tanis in Egypt at 31°.[93] It was found in the private house of perhaps an official who worked at the nearby temple of Amun. Fixed in place, the dial 'may have kept the owner abreast of time in between his duties at the temple, averting tardiness whenever he had to return to them'.[94] It seems that not all portable dials were intended to be taken far.

Dating slightly later to the first century AD is a portable sundial from Herculaneum. Known as the 'Ham Dial' because its distorted bronze plate looks just like a small leg of ham, it consists of a spike on one side, which threw a shadow on to a series of crisscrossing lines on the plate, from which one could read the hour of the day.[95] Another early portable dial also dating to the first century AD is a tiny Roman cylinder-type made of bone and bronze, from near Este in northern Italy.[96] Other portable bronze dials are circular in shape, and some come with extra plates to suit different latitudes.[97] One small dial of perhaps mid-third century AD date consists of just two plates and a *gnōmōn* and yet permits the reading of the time of day anywhere between latitudes 30° and 60° N; thirty locations are specifically listed.[98]

These dials foreshadow the astrolabe, a portable timepiece characteristic of the Middle Ages (Figure 4.7).[99] None survives from antiquity, but the instrument was probably a late Greek invention. Via a system of stereographic projection the celestial hemisphere was represented in two-dimensional form. Ptolemy describes this method of representation in his *Planisphere* in the second century AD and it may have been known to Hipparkhos in the second century BC. The earliest surviving description of the astrolabe, by

Figure 4.7 Istanbul, Archaeological Museum, inv. 2970: Arab astrolabe, signed 'Hasan bin Ömer', AH 681 (AD 1283).

Source: Photograph R. Hannah.

Philoponos in the sixth century AD, depends on an earlier one by Theon in the fourth century AD.[100]

To what extent any of the principal kinds of sundial were related in an evolutionary fashion is debatable. It has been thought, for instance, that sundials progressed from human shadows, through formal shadow tables, to spherical sundials, to plane dials.[101] The notion that spherical sundials constituted the earliest constructed type is based on the assumption that the *polos*, mentioned as early as Herodotos (2.109), was a sundial which in its form mirrored the celestial dome. But others have seen no evolutionary development from one type to another, because of Vitruvius' inventor-list and on the basis of the actual dials which have survived. Gibbs, for instance, argued that if one used the dates of the inventors listed by Vitruvius as an indication of the developmental sequence of the various types of sundials, then several types (the horizontal, meridian and hemispherical) were invented simultaneously, followed by the cut spherical, then by deviating and conical dials, with the roofed sundial being probably one of the latest. She found no obvious development or modification of a type once it had been invented and a construction process for it established.[102] More recently, Schaldach has found no development from the spherical dial to other forms and is indeed inclined, on the basis of the earliest surviving Greek sundial – the plane dial from Oropos – to regard the *polos* as a plane sundial type too.[103] My own view at present is that there is apparently a formal link between human shadow 'dials', shadow tables and horizontal plane dials, sufficient to warrant hypothesising a developmental link between them; and that a similar formal similarity between the celestial dome, celestial globes and spherical sundials supports the supposition of a developmental connection between these, stemming perhaps from Eudoxos himself, as Turner proposed.[104] The gaps in the archaeological record may in time be filled by future discoveries or excavations, as the plane sundial from Oropos warns us. In this case, a dial of unexpectedly high sophistication at an early date signals complicated, prior developments, for which we have as yet no further material evidence, but which we remain obliged to assume.[105]

In both lines of presumed evolution, the concepts of time measured by the different types of sundial derive from human perceptions of the sun's apparent movement in the sky over the year. This movement is measured from the shadows cast by *gnōmōns* on to different forms of receiving surface. What evolves is a sense of standardised time-reckoning, as the shadows are constrained ultimately by use of an artificial, non-human *gnōmōn*, and then parcelled into periods of time (zodiacal months), which are themselves based on abstractions of reality (zodiacal signs of standard length as opposed to zodiacal constellations of varying length). Both lines of development through flat and spherical surfaces to receive the *gnōmōn*'s shadow are arguably of eastern origin, though there is nothing to prevent both being indigenous to Greece, and simply resembling the foreign responses to the same phenomena.

5

MEASURING TIME

It remains characteristic of ancient time technology that stone sundials were, by and large, not used to tell daily calendar time. Even the largest, the conical specimens from above the Theatre of Dionysos in Athens and from the sanctuary of Apollo at Klaros in Turkey, which present meridian lines of 480 mm and 397 mm in length respectively and so would have the capacity to demarcate small periods of time, still notify the observer only of the solstices and equinoxes within the year; there is no interest in displaying the zodiacal months, let alone the days within them.[1] As for time within the day, the smallest diurnal unit on surviving sundials is the half-hour, and that is very rare: neither of the two large dials just mentioned bothers to display it. Yet occasion did demand the *measurement* of much smaller units of time within a day, and this need drove the development of other forms of time-keeping technology.

So far we have studied means of marking time which were based in one way or another on the perceived motions of the celestial bodies: observations either of the motions themselves, particularly their rising and setting over the horizon, or of reflections of those motions via shadows. With the devices which measured periods of time, however, we find ourselves dealing generally with a different perception of time: time that can be divided into sections which may be equal throughout the year, regardless of the season.

Various devices were used to measure small pockets of time. For example, Pliny (*Natural History* 33. 96–7) mentions how men in a deep Spanish silver mine baled out the water in shifts measured by lamps.[2] The ordinary, ubiquitous, clay oil lamp lent itself admirably to this function, since it could burn for a determinable period of time, depending on the capacity of the bowl, the material of the wick, the size of the flame, and the type of oil.[3] Most of these variables feature in the instructions for the use of lamps in magical rituals, such as conjuring up and holding on to gods or spirits for their favours, preserved in Greek papyri from Egypt. The following description of a ritual to summon the god Apollo gives the flavour of the type without any particular reference to time:[4]

> Invocation for Apollo:
> ... Taking an unreddened lamp, prepare it with a strip of linen and rose oil or nard oil, and clothing yourself in an oracle's garment, hold an ebony wand in your left hand and the amulet in your right. ... When you have completed everything aforementioned, call him with this spell: ... And after the enquiry, if you wish to release the god himself, transfer the aforementioned ebony staff, which you have in your left hand, to your right, and transfer the twig of laurel, which you have in your right hand, to your left, and put out the burning lamp ...
>
> (*PGM* I. 263–347[5])

But time could matter, especially when conjuring up powerful spirits, and the lamps provided a handy and cheap way to keep an eye on the time. The development of regional types of mould-made lamp with fixed dimensions seems to have generated an awareness of more or less fixed periods of time per lamp-type. The lamps in the magical rituals, for instance, were sometimes intended to hold fixed amounts of oil according to the spell, thus setting different lengths of time in which the conjuring spell could be effectively or safely performed:

> Then, having come to the day, in the middle of the night at the fifth hour, when it is quiet, having lit the altar, have nearby the two cocks and the two lamps, lit – let the lamps be of quarter volume – to which you add no more oil. ...
>
> (*PGM* XIII. 122–7[6])

A similar mentality towards measuring time is encountered much later in the medieval period in northern Europe, where the oil lamp was less common, because the fuel was not native, and was superseded by the wax candle. The apocryphal story is told of Alfred the Great in the ninth century inventing the lantern clock, which comprised six candles which burned (presumably one after another) through a whole day and night. Each candle would therefore last four hours. They were also each marked off into twelve sections, which would suggest that each division was meant to represent the passage of 20 minutes (four hours = 240 minutes, divided by 12 = 20 minutes).[7]

In both cases – clay oil lamps and wax candles – time was measured via the consumption of controllable volumes of flammable fuel. The medieval sand glass would afford a further device for the same purpose, safer to handle but harder to manufacture.[8] All mechanisms provided the facility to measure the passage of time within the day or night. In that sense, they achieved the same effect as a sundial, but overcame the problems that arose from the lack of sunlight owing to inclement weather or night-time. But unlike a sundial, lamps, candles and sand glasses fail to tell the time: they cannot in themselves indicate either the time of day as such-and-such an hour, or the period of the year one happens to be in. All are simply timers.

As timers, these devices provided an even measure of whatever division of the day was desired. Each unit of measure was the same as the preceding one and the next one. This is an interesting point, insofar as with sundials we became used to the idea of seasonally variable units of time (hours), longer in summer, equal at the equinox, but shorter by winter. In contrast, from the Hellenistic period onwards astronomers regularly used equinoctial hours in their calculations. Yet only one sundial, the early plane dial from Oropos, currently demonstrates the facility to measure equal, or equinoctial, hours regardless of the season.[9] So the Greeks and Romans were plainly capable of thinking in quite different terms when it came to measuring small blocks of time within the day. The puzzle is why they chose not to standardise their time-divisions on sundials into equal segments in public, non-scientific contexts. If the equinoctial hour was utilised in sundial technology so early on at rural Oropos, why did it not become the hour-type of choice through the Hellenistic and Roman periods in the city centres?

It may be that the geometrical and astronomical knowledge required to construct the Oropos sundial was more than most manufacturers could assimilate – and this one seems to betray signs of hesitation on one face of the dial. Equatorial plane dials are not at all common in the surviving corpus. The equatorial cylindrical dial discovered in Ai Khanoum is unique, and even it did not use equal hours. The answer to the question of the relative popularity of unequal hours must lie to a large degree in the relative popularity of the spherical and conical types of sundial in classical antiquity. These present dial faces which more or less mirror the perceived celestial dome, a characteristic which may have given the forms their preferred status among makers and users. If one wanted to subdivide daytime on these types of dials through the day, the very geometry that underlies the forms of the dial faces demands, in practical terms, subdivisions which create unequal hours from one day to the next.

One other major form of timer existed in antiquity: this was the water clock. Initially it too, like the lamp, candle or sand glass, did not derive its units of measure from mimicry of the motions of the celestial bodies, yet even it was drawn eventually into matching its measures with the heavens. The earliest surviving example is Egyptian, dating to about 1400 BC, while the earliest mention of such an instrument there is from the late sixteenth century BC, but the Egyptians may well have borrowed the device from the Babylonians. Egyptian models were bucket-like containers, which did in fact seek to measure time through the day and the year, but in so doing had to cope with several problems: varying outflow rates owing to lower pressure as the level fell; the use of unequal, seasonal hours; variations in temperature both between day and night and between seasons, which affect the viscosity of the water and hence its rate of flow; and the quality of the water, as silt would soon clog up the outflow hole. The Egyptians managed these problems by tapering the water clocks to the bottom in an attempt to equalise the flow of water as the level fell; by marking on the interior a series of different hour-scales, one for each month of the year; and probably by sourcing their water carefully and allowing the silt to settle out before pouring it into the clock.[10]

The Babylonians too, to judge from a very late source, had water clocks of a similar bucket-like kind. Sextus Empiricus, writing in the second century AD, explains how the 'Chaldaeans' (a common term for the ancient Babylonians) sought to divide the circle of the zodiacal stars into twelve parts:

> . . . after watching closely a bright star in the zodiacal circle rising, the ancients, having filled a perforated amphora with water, then let it flow into another vessel set underneath until the same star rose again, guessing that the turning of the circle was from the same sign to the same sign, they then took the twelfth of the flow and considered in how much time it had flowed; for they said that in that time also the twelfth part of the circle had come back, and the aforesaid part of the circle has to the whole circle the same ratio as the part of the water which has flowed has to the whole of the water. From this reference-point – that of the twelfth part, I mean – they marked off the final limit from some conspicuous star observed at the time or from one of the more northerly or southerly stars which rise at the same time. They did the same also for the other twelfths.
>
> (Sextus Empiricus, *Against the Professors* 5.24–6)

In spite of the great antiquity of these vessels and Greece's familiarity with eastern cultures from an early stage, the Egyptian-style water clock does not appear to have influenced Greek developments of such instruments until the Hellenistic period.[11] In Greece what is nowadays called a water clock was originally not a clock or calendar at all, but just a simple timer, which measured the set times for certain activities by the fall or rise in water level in a container. The name given to the instrument was *klepsydra* ('water thief'), a term apparently borrowed from a device which worked like a large, bulbous pipette (e.g. Aristotle, *On the Heavens* 294b.14–30; [Aristotle], *Problems* 914b).[12] The same name, interestingly, was applied also to a stream which ran, mostly underground, from the north slope of the Akropolis in Athens and probably fed one of the most famous water clocks of antiquity, the Tower of the Winds.

The sixth-century AD Neoplatonist, Simplicius, in his commentary on Aristotle's *On the Heavens*, usefully explains the form and function of the pipette-like instrument for his readers:

> The *klepsydra* is a narrow-mouthed vessel with a wider base perforated with small holes, what they now call a *hydrarpax*. When this vessel is lowered into water while the mouth at the top is stopped, the water does not go in through the holes, because the massed air in it resists the water and prevents its entering upwards from its inability to change its place. When the stopper of the mouth at the top has been removed, the water goes in, as the air makes way for it.
>
> (Simplicius, *On Aristotle's On the Heavens*, vol. 7, page 524, lines 19–25[13])

So, when put into water with its upper mouth open, the *klepsydra* would fill up by 'stealing' the water and would then sink. The device was used experimentally to demonstrate the qualities of air through its displacement by water.

One imagines this form of *klepsydra* to be quite small. It is also essentially an inflow device, at least in the hands of the physicists. It could also be used as an outflow type, if the mouth was bunged once the vessel was full or partly full of water and it was then lifted from the container of water and the bung removed.[14] But other types of outflow timing devices existed, of quite different form from the 'pipette' variety, and some of them very large indeed.[15] The name *klepsydra* was simultaneously applied to these different types of containers, which were used at least from the fifth century BC so as to time non-scientific activities, notably speeches in the law courts. Indeed, the *klepsydra* became synonymous with the courts, as Aristophanes' plays show (Aristophanes, *Acharnians* 694, *Wasps* 93). But more than this, by the mid-fourth century BC the orator Demosthenes could use the word 'water' (*hydōr*) as a synonym for 'time' itself. He talks of the period allotted for his speech in the law court as being his 'water' (*Against Polykles* 2, *On the Crown* 139, *On the False Embassy* 57).

Different measures were used for different speeches. Ten *choes*, for example, were made available for cases involving more than 5,000 *drachmai* and seven *choes* for those under that amount ([Aristotle], *The Athenian Constitution* 67). In his defence speech *On the False Embassy*, Aiskhines tells of having eleven amphoras allotted him (Aiskhines, *On the False Embassy* 126). In telling the story of a case about a right of inheritance, Demosthenes complains of being given only a quarter of the time that the claimants were granted – to their one amphora he has just three *choes* (Demosthenes, *Against Makartatos* 8). An example of a judicial *klepsydra* has been excavated in the Agora in Athens, dating to ca. 400 BC.[16] A bucket-like vase with a piped hole near the base for the outflow and marked on the outside with the Greek letter *chi* twice (XX, i.e. two *choes*), it was found by experiment to hold 6.4 litres or the equivalent of just six minutes' worth of water, a refinement of time unattainable by sundial.[17] The *klepsydra* was treated much like a modern stopwatch.[18] Demosthenes requests of the court official, 'Stop the water!', when the taking of depositions interrupts his speech (*Against Euboulides* 21, *Against Konon* 36, *Against Stephanos* I. 8). Dramatic plays too were apparently timed before Aristotle's time by the *klepsydra*, in this case a container presumably on a larger scale (Aristotle, *Poetics* 1451a7–9). Even sessions with an *hetaira* (a prostitute) were so timed, Athenaios (13. 567c–d) tells us, referring to a play by Euboulos, whose main character, an *hetaira* called Metikhe, gained her nickname, Klepsydra, from timing her sessions with her clients until the *klepsydra* was emptied.[19] Perhaps there was also a visual gag in this last usage, if the form of the *klepsydra* resembled male genitalia, as the bulbous 'pipette' and the piped bucket both could.

When such a timing device was introduced into Athens is unknown. If it came from one of her Eastern neighbours – Egypt or Babylon – then it is strange that it does not make an earlier appearance than the late fifth century, since Greek contacts with both areas stretch back a considerable time. One guess is that since Herodotos does not mention anything like a water clock in his history of Greece down to the Persian Wars early in the fifth century BC, and yet by his own time towards the end of that same century it has become commonplace in the law courts, then the device may have been brought to Greece in the intervening period, and therefore in the course or wake of the wars with Persia, or through Ionian Greek contacts with the nearby Persian Empire.[20] As we have seen, there is some limited testimony that the Babylonians used such an instrument as a timer, and it is as a timer, not a clock, that it appears in Athens.

It may be that the introduction of the *klepsydra* to civic contexts like the law courts signifies an increasing consciousness of time from the late fifth century BC, coinciding as it does with Meton's setting up of a *hēliotropion* on the Pnyx, near the Assembly area.[21] In both the law courts and the Assembly there was a tradition of beginning proceedings at daybreak, forewarned by cock-crow or other signals (Aristophanes, *Acharnians* 19–20, *Ekklesiazousai*, 30–1, 82–5, 289–92, 390–1; *Thesmophoriazusai* 277–8, *Wasps*, 689–90). From the late fifth century, however, we find artificial mechanisms being introduced: not clocks as such, though the fourth-century sundial at the Amphiareion at Oropos warns us that their time will come soon, but timers in the form of the *klepsydra*. What caused people to look for such instruments can only be conjectured. It may be that the burgeoning democracy in Athens, with its extremely high degree of public accountability for its officials, led to a steep increase in litigation, and so provided a fertile breeding ground for ideas on how to keep political life running smoothly by constraining activities within certain time limits.

Yet if such a need for timekeeping was felt, the aversion to being controlled by a time machine which we have already encountered with sundials is found also with the water clock: Plato has Sokrates express his belief that in contrast to the freedom to expound at length and at will enjoyed by philosophers, the pressing needs of the law court, including the flow of water from the *klepsydra*, enslave lawyers (*Theaetetus* 172c–e). But while the necessity for brevity imposed by the *klepsydra* might on occasion have diverted contestants in the courts from the truth and led to fallible judgments,[22] it also ensured equality through standardisation, a fundamental characteristic of the democracy. In this case, it was time that was standardised, but in the second half of the fifth century BC Athens also sought to make uniform the coinage, weights and measures used within its wide sphere of influence.[23]

The *klepsydra* could also be used to jolly along the democratic process. At Iasos in south-west Turkey in the third century BC a decree was passed which, among other things, established that a clay-pot of fixed size (a *metrētēs* by volume) should be set up high enough for everyone to see it, and made to have its water run out from sunrise. Citizens who failed to get to Assembly before the water had run out then forfeited their attendance fee.[24] Those citizens in modern democracies who face a legal requirement to vote at elections may recognise the 'encouragement', though its form is unusual.

The standardisation of time through the *klepsydra* extended also to the whole of the legal day in Athens, in those cases that demanded more time.[25] This so-called 'measured day' was made to correspond, regardless of the time of year, to the length of the shortest days of the year, those of the Athenian month Poseideōn in midwinter ([Aristotle], *Athenian Constitution* 67.4; Harpokration, s.v. *hēmera diamemetrēmenē*), or about nine-and-a-half hours in our terms.[26] This was then subdivided into a certain quantity of water – the fourth-century orator Aiskhines (*On the False Embassy* 126) talks of one such day being equated with 'eleven amphoras' – with thirds being given to the prosecution, the defence and the judgement. Once again, it is interesting to note how time is restricted presumably in the interests of equality: access to the same amount of time for court cases throughout the year was assured by correspondence with the shortest day of the year, ensuring that cases tried in every season would still occupy only the daylight hours. The equinoctial and the solstitial days, for instance, would have been too long in winter.

Despite Sokrates' avowed dislike of the tyranny of the water clock, Athenaios (4.174) tells of Plato himself making a *nukterinon hōrologion*, a night-time clock, which was 'like a very large *klepsydra*', but what sort of *klepsydra* was intended is impossible to recover, and so, therefore, are the device's workings, although this has not prevented modern reconstructions being promoted.[27] The use of the term *hōrologion* for such an instrument may reflect usage in later times rather than Plato's, and any suggestion that it worked hydraulically to emit a sound through a whistle and so served as an alarm clock, as some modern commentators have suggested, is probably anachronistic, reflective of technology that smacks more of Hellenistic Alexandria than of Classical Athens.[28]

Hydrologion is another term applied to outflow water clocks, and we find it used in scientific circles, for instance, in an experiment to deduce the size of the sun relative to its full circuit of the sky over a day and night:

> For when it [the sun] is measured with *hydrologia*, it is discovered to be 1/750th part of its own circle. For if, let us say, a *kyathos* flows in the time the sun rises fully from the horizon, the water let to flow out in the whole day and night is discovered to be 750 *kuathoi*. Such an approach is said to have been invented first by the Egyptians.
>
> (Kleomedes, *The Heavens* II.1.184–91[29])

Kleomedes is here reporting a method outmoded among astronomers by his own time in the first or second century AD, as it had already been criticised by Hipparkhos in the second century BC, according to Pappos (ed. Rome, I. 87–9), and is rejected later by Ptolemy (*Almagest* 6.14) around the time of Kleomedes for its lack of accuracy in comparison with readings provided by the *dioptra* described by Hipparkhos.[30] Of course, even without measuring the time by some such mechanism between sunrise and sunset, people were well aware that the lengths of day or night differed in different latitudes. According to Krates of Mallos, a philosopher of the second century BC, Homer demonstrated as much in his description of the fantasy land of the Laistrygones:

> There a sleepless man earns double wages, one grazing cattle, the other tending silver-shining sheep; for the roads of night and day are close together.
>
> (*Odyssey* 10.84–6)

According to Geminos (6.10–11), Krates interpreted this passage scientifically as a reference to a place in a far northern latitude, where the longest day could be 23 equinoctial hours.[31]

However rough and ready the calculations provided by water clocks were for sophisticated astronomical purposes, they continued in use in scientific, political, legal and military contexts. In Rome, Scipio Nasica was credited with introducing the first water clock, in 158 BC, just a few years after the first accurate sundial was set up in the Forum (Pliny, *Natural History* 7.214). The *klepsydra* divided the hours of the day and night equally, Pliny records, and it can therefore be classed, along with its nearby sundial, as a clock, allowing the time to be told through the day. Yet the earlier function of the *klepsydra* as a timer was not lost on the Romans. We see this clearly in the law courts.

In the Roman law courts a speech for the prosecution or defence was timed against the water clock. In the latter part of the first century AD, the satirist Martial could berate a lawyer for the length of his speeches:

> Seven water clocks, Caecilianus, the judge reluctantly gave you, when you demanded them with your great voice. But you speak long and loud, and with your head thrown back you drink warm water from glass cups. So that you may satisfy both your voice and your

> thirst, Caecilianus, we suggest that from now on you drink from the water clock.
>
> (Martial 6.35)

The lawyer consumes with his speeches the time that seven water clocks take to empty. Martial recommends that instead of quenching his resultant thirst from glasses of water, the advocate should simply consume the contents of the water clock directly, and thereby (he implies) shorten the proceedings. How long Caecilianus took to orate may be calculated from an instance given by his younger contemporary, Pliny the Younger, a lawyer himself. Pliny relates that in one very important court case, 'I spoke for almost five hours, since to the twelve water clocks of the largest size that I received, four were added' (*Letters* 2.11.14). From this description we can calculate the rate of outflow: with sixteen very large water clocks providing almost five hours' speaking time, one water clock was worth about half an hour. So Martial's loquacious lawyer had orated for about two-and-a-quarter hours, if his water clocks were of the same size as Pliny's.[32]

The length of Pliny's speech may be compared with what was permitted under a law passed in 44 BC for Julius Caesar's colony at Urso (modern Osuna) in southern Spain. This stated:

> No *duovir*, who holds a *quaestio* or administers a trial in accordance with this law, if the trial is not stipulated by this law to take place in one day, shall hold the *quaestio* or administer the trial before the <first> hour or after the eleventh hour of the day. And the said *duovir*, for each accuser, shall grant the opportunity of making the accusation for four hours to that one of them who shall be the chief prosecutor, and for two hours to the one who shall be the assistant prosecutor. . . . For as many hours in all as it is appropriate for all the accusers to be granted the opportunity of speaking in each *actio*, he is to grant to the defendant or to whoever shall speak for him so many hours and as much again in each *actio*.
>
> (Crawford 1996: 409 no. 25, ch. 102[33])

Here a fixed amount of time, six hours in all, is allotted for the prosecuting party, and double that time for the defendant. In principle, the chief prosecutor was allowed four hours to speak, while an assistant prosecutor was granted a further two hours, but one could transfer some of his allotment to the other, while still having to stay within the maximum six hours for their part of the proceedings. The action, it may be noted, was to take place between the first and eleventh hours, and therefore only in daylight, so the court no doubt ran water clocks to time the speeches, but will have relied on sundials or different water clock devices to signal the period of the day within which the cases could be heard.[34]

Other cases, however, could be more constrained. A law passed in 52 BC by Pompey the Great (Asconius, *For Milo* 31) imposed a strict limit on the length of legal proceedings and speeches in cases involving violence and electoral bribery. The first three days were devoted to hearing testimony and taking depositions, and then on the fourth day, after the jury panel had been reduced, the speeches were heard, the prosecutor being restricted to just two hours, and the defendant to three.[35] Thereby, complained Tacitus (*Dialogue on Orators* 38.2), Pompey put 'as it were, reins on eloquence'. It might be said, however, that Tacitus, a contemporary of both Martial and Pliny the Younger, was not a disinterested party, as he was himself a lawyer. It is clear that Pompey's restrictions separated the speeches from the taking of depositions, thus preventing the advocates' speeches from being interrupted as Demosthenes' had been in fourth-century Athens, when he had to ask the official to 'stop the water' while the depositions were taken. What sort of environment the law of Pompey was situated in may be gauged by looking at the speeches of his contemporary, Cicero. A recent recreation of Cicero's speech in defence of Archias the poet (62 BC), declaimed in Latin (and toga, one might add), shows that it would take thirty-five minutes to deliver.[36] But this is a very short speech for Cicero. His defence of Sestius (56 BC), for example, if delivered at the same rate as that for Archias, would take three hours and eighteen minutes to present. It was soon afterwards, in 52 BC, that Pompey had his law passed. It was remembered later by Tacitus, who saw it as restrictive on the practice of rhetoric, and yet, as the instances of the statute at Urso and the case of Pliny the Younger demonstrate, the Pompeian law was not universally applied to all types of law suits.[37]

Lengthy proceedings can, of course, lead to another form of clock-watching. Bored participants in the law courts could send their slave boys off elsewhere to find out what hour it was (Cicero, *Brutus* 200). The practice was well known in everyday life, with the wealthy able to have a slave check outdoors for the time of day from a sundial (Martial 8.67.1, Juvenal 10.216). Seneca complains about the leisured élite who have to be reminded by someone else 'when they should wash, when to swim, when to dine' (*On the Shortness of Life* 12.6). But Pliny the Younger describes with admiration the daily regimen of his older contemporary, the elderly Spurinna, who has his bath hour announced to him (Pliny the Younger, *Letters* 3.1). Pliny's uncle lists among those fortunate to die suddenly the ex-praetor Gnaeus Baebius Tamphilus, who died after asking his slave boy for the time (Pliny, *Natural History* 7.182); clearly, his time had finally come. Another individual, a lawyer in Dalmatia, very unusually, even had the time of his death noted on his tombstone, down to the very hour:

> Sacred to the spirits of the dead, Q(uintus) Publicius Aemilianus,

> lawyer, by nationality African, lived 47 years, 9 months, 7 days, in the fifth hour of the night.
>
> (*CIL* 3.2127A; *ILS* 7774[38])

The time must have been taken from a water clock. The hour itself could be a killer: in the midst of an early spring day, Horace morbidly reminds us: 'The year warns you, lest you hope for immortality, as does the hour, which seizes the nourishing day' (Horace *Ode* 4.7.7–8).[39] Dancing with death, on the other hand, was Trimalchio, Petronius' fictitious rich man, who has a *hōrologium* in his dining room – so necessarily, again, a water clock – and a trumpeter 'so that he could know at once how much of his life he had wasted' (Petronius, *Satyricon* 26.9).

This dependence on another to find out the hour of day was the emperor Domitian's undoing: forewarned that the fifth hour was to be the hour of his death, he was falsely put at ease at the fifth hour on the day of his assassination by being told that the time was the sixth hour, and so, thinking the deadly hour had passed, he was off his guard when the attack came (Suetonius, *Domitian* 16.2). Yet, ironically, Seneca could complain that sundials were notoriously inaccurate: the hour of the emperor Claudius' death cannot be clearly identified because 'it is easier to find agreement among philosophers than among sundials', and so he opts for somewhere 'between the sixth and seventh hours' (*Apocolocyntosis* 2).[40]

Julius Caesar, in a mixture of scientific curiosity and military need, measured the length of the nights in Britain 'by water', presumably with *klepsydrai* (*Gallic War* 5.13.4), and the popular tables of the lengths of day or night in terms of hours on the longest day reported by, for instance, Geminos (6.7–8), Strabo (2.5.38–42), Pliny the Elder (*Natural History* 2.186) and Kleomedes (II.1.438–51), may still reflect the results of this older technology. Kleomedes' system for calculating the monthly increase in daylight from the shortest day to the longest (Kleomedes I.4.18–29), while smacking of an arithmetical schema, is nonetheless reasonably accurate.[41] One wonders therefore to what extent it was based on empirical data derived from measurements taken by water clock.

While relatively high-speed travel in the nineteenth century gave the impetus to states to establish national standard times and international time zones, no such pressure was felt in the ancient world because there was no equivalent to the steam train. Interestingly, the other technological development of the nineteenth century which drove a need for standardised times – the telegraph[42] – did have a parallel, if not a direct ancestor, in antiquity, but it seems also not to have increased awareness of regional time differences, but instead to have fed a desire simply to conquer the tyranny of distance and its braking effect on the speed and accuracy of communication. This parallel was the fire signal, used to send messages across considerable distances more quickly than a rider or sailor could travel to convey the same message.[43] The technology is famous from its dramatic use at the start of Aiskhylos' *Agamemnon* (1–39), where news of the fall of Troy is sent by fire beacons across the northern Aegean, down through mainland Greece, and so to Mycenae in the Argolid (Klutaimestra gives the full itinerary at *Agamemnon* 281–316). Polybios, three hundred years after Aiskhylos, still extols the virtues of fire signals across both space and time:

> Now, that timing plays a major role in all enterprises, but especially in military matters, is clear to everyone; and of those things that help towards this, fire signals are the most powerful. For whoever has an interest can know both what has just happened, and what is still in progress, sometimes if he is three or four days' journey away, sometimes even further. Thus always in those circumstances where assistance is needed, there is unexpected help because of the message from the fire signals.
>
> (Polybios, *Histories* 10.43.2–4)

But there were problems with such signalling, such as how to convey unexpected or complex messages (Polybios, *Histories* 10.43.8). In an attempt to circumvent these difficulties, Polybios tells us, Aineias the Tactician in the mid-fourth century devised a mechanism combining the timer-facility of water clocks with the visibility of fire signals (Polybios, *Histories* 10.44). Aineias advises setting up a pair of large clay tubs (3 cubits in diameter and 1 in height, or about 1.35 m by 45 cm), fitted with a tap and a cork float pierced by a graduated rod, which was to be inscribed with various messages indicating events likely to occur in battle, such as 'cavalry have entered the country', then in successive sections 'heavy foot soldiers', 'light-armed soldiers', 'foot soldiers with cavalry', 'ships', 'corn', and so on. The tubs would be tested to ensure that they emptied at exactly the same rate, and then set up in two different places within view of each other. When a particular event then takes place in the battle, which has been anticipated in the messages, a fire signal from the battle-site is lit and answered by one from the other, and at both sites the tubs' bungs are then pulled out, letting the water flow out down to the point where the relevant message can be read on the first tub. Another fire signal is lit at the battle-site, and at the second site the same message should be able to be read and appropriate action taken to send assistance. Polybios himself expresses doubts about the efficacy of this system, mentioning some problems – most obviously, could the water clocks anticipate all possible eventualities in their messages? – and the description leaves a lot to be desired in terms of detail and clarity.[44] Kleoxenos and Demokleitos devised a different system, akin to semaphore, to which Polybios says he himself added improvements, preserving and indeed increasing the torches, but removing the water clocks and instead adding tablets inscribed with the letters of the alphabet, which were to be held up like semaphore flags to spell out words, and which were to be viewed through a telescope-like *dioptra*. This system was intended to remove the imprecision and inflexibility of the previous water clock messages (Polybios, *Histories* 10.45.6–47.2). In terms of time technology, however, it uses only the torches, still to signal to the parties when to start sending messages.

Klepsydrai of some kind were also used to measure out the length of night watches in the military world. But there a measure based on subdivisions of a standardised, winter day, such as Athens instituted in its legal 'day', could not work for the longer days of the campaigning season, so attempts were made to ensure that the watches, and hence the responsibilities, were equalised by adjusting the volume of water that the clocks held through coating the inner surface of the *klepsydra* with varying layers of wax:

> The way that the watches may fall equally and on all, as the nights become longer or shorter, is to keep watch <entirely> by water clock, changing the clock every ten days. Preferably its inside should be waxed, and the wax should be reduced so that more water can be held when the nights are lengthening, and increased, so that it holds less, when they are shortening. Let that be enough explanation from me about the equalisation of the watches.
>
> (Aineias the Tactician 22.24–5)

This is an awkward testimony, as the transmitted text is unreliable, and we do not know what sort of *klepsydra* Aineias had in mind. The ten-day rule exists only on the basis of emendation of the text, but seems to recall the ten-day change of decanal stars that the Egyptians adopted to denote the hours through the night;[45] that, however, would lead to different hourly scales for the changing seasons, not different layers of wax. Certainly it is interesting that Aineias seems otherwise to have been unaware of the Egyptian form of water clock, an unfamiliarity which gives some support to the notion that the seasonal hours were not introduced to Greece until later in the fourth century BC and then via Egyptian water clocks.[46] Young thought an open vessel would have been required to suit Aineias' instructions for layering it with wax.[47] West took it that the vessel would also need to be larger than the Agora *klepsydra*.[48] Pattenden, on the other hand, took Aineias' *klepsydra* to be of the early 'pipette' variety, with holes in the bulb able to be bunged up with wax to vary the volume of water that could flow into the vessel, thus causing it to sink at varying times after immersion.[49] The problem with this suggestion, as Whitehead has pointed out, is that it would invalidate Aineias' procedure of layering the wax: removing bungs of wax would cause the *klepsydra* to sink more quickly, which is what would be wanted on shorter, not longer, nights. Instead Whitehead has reverted to larger, open vessels, perhaps of the kind found near the Heliaia in the Agora in Athens or that found at the Amphiareion in Oropos from the second half of the fourth century BC, and so contemporary with Aineias; Lewis supports this conjecture.[50]

These last two *klepsydrai*, from the Athenian Agora and the Amphiareion at Oropos, are very large outflow water clocks with a capacity of about 1,000 litres. They are sufficiently close in form as to suggest that the same person may have designed both. The Oropos example is the better preserved, and calculation of its outflow suggests that both specimens would take about 17 hours to empty, long enough to operate uninterruptedly over the whole of a summer's day of under 15 hours.[51] At Oropos the clock may have been used to time rituals or performances at the nearby theatre.[52]

The Athenian *klepsydra*, however, showed signs of being redesigned from an outflow to an inflow type over the period of its use down to the first half of the second century BC. In its first version, as with its counterpart at Oropos, the clock was filled at the start of the day, and then allowed to empty gradually through the day or through a more limited period if desired. A gauge of some kind was presumably fitted to enable people to see how much water had passed, and therefore how long a period of time had elapsed or was left to elapse for whatever activity was being measured.[53] This outflow was regulated only by the narrowness of the outlet pipe at the bottom of the tank. Unfortunately, this pipe did not survive the later renovation at Athens, and its hole was filled in; but the equivalent at Oropos has survived, and comprises a bronze cylinder, 9 cm in diameter, terminating in a hemisphere, which is punctured by a hole of only 2.6 mm diameter. Water would therefore exit the tank at a slow rate through so narrow an outlet. But if the tank had vertical walls, then as it emptied, the pressure would decrease and the rate of exit for the water would slow down. Any gauge against which the outflow was to be measured ought to take this rate of change into account, in the interests of equity, with unequal calibrations for the periods of time being measured, but it is not clear that this was possible.[54] As we have seen, the Egyptian solution to the problem, on a much smaller scale, was to slope the walls of the conical, bucket-like container, so that the pressure could be equalised as far as possible; a limestone *klepsydra* from Karnak, dated to about 1400 BC, has slopes set at 70° to the horizontal, and achieved a nearly uniform rate of emission.[55] At Oropos the upper part of the walls of the tank were sloped outwards and then the rest of the tank was made vertical, so as to produce roughly equivalent rates of emission through the day.[56]

An alternative method was to provide a constant source of water and to insert an overflow pipe near the top of the *klepsydra*, so that the level of the water and its pressure remained constant. This became the basis of the inflow type of water clock.[57] A version of this method was introduced for the Athenian water clock in its later renovation. Two reservoir tanks were added above the main tank (one for supply, the other perhaps for overflow), an overflow pipe was apparently added on the outside of the tank (but does not survive), and the first outlet pipe at the bottom of the tank was filled in and replaced with another larger drain hole of 16 mm diameter, which allowed a much more rapid exit of water in less than 40 minutes. With the outlet now so open as to make it useless for an outflow water clock, the tank must have been transformed into an inflow type, by which the time was measured not by falling water in the main tank but instead by the rising water level in it.[58]

A float, whether in the outflow tank or in the inflow, could be connected via gear wheels and even automated figures to some means of displaying the passage of time, as water entered or exited the tank. This time display could be a simple cylinder marked with a scale of hour lines. The difficulties presented by the need to represent the uneven seasonal hours persisted, yet however retarding to further mechanical development these hours may seem, Greek ingenuity found ways to overcome them. Vitruvius (*On Architecture* 9.8.4–15) describes sophisticated examples of such water clocks devised by Ktesibios in the third century BC.[59] Later in the same century Arkhimedes devised further refinements, which improved the engineering and hence accuracy.[60]

An alternative approach to calibrating time against the seasonal hours was to tie the passage of time explicitly to the stars. This development brings us back to measuring time via the motions of the celestial bodies, but in a mechanism that began life completely divorced from the heavens. It demonstrates the strength of the attraction for the ancients towards keeping time in line with the cosmos. The end-result is the anaphoric clock, which told the time via an automated representation of the sequential risings of the stars represented by images of the zodiac or through stereographic projections of the celestial sphere.[61] Vitruvius (*On Architecture* 9.8.8–15) is our principal source for the description of this form of *klepsydra*:[62]

> With the aid of the analemma the hours are marked by brazen rods on their face, beginning from the centre, whereon circles are drawn, shewing the limits of the months. Behind these rods a wheel is placed, on which are measured and painted the heavens and the zodiac with the figures of the twelve celestial signs, by drawing lines from the centre, which mark the greater and smaller spaces of each sign. On the back part of the middle of the wheel is fixed a revolving axis, round which a pliable brass chain is coiled, at one of whose ends a phellos or tympanum hangs, which is raised by the water, and at the other end a counterpoise of sand equal to the weight of the phellos.
>
> Thus as the phellos ascends by the action of the water, the counterpoise of sand descends and turns the axis, as does that the wheel, whose rotation causes at times the greater part of the circle of the zodiac to be in motion, and at other times the smaller; thus adjusting the hours to the seasons. Moreover in the sign of each month are as

many holes as there are days in it, and the index which in dials is generally a representation of the sun, shews the spaces of the hours; and whilst passing from one hole to another, it completes the period of the month.

Wherefore, as the sun passing through the signs, lengthens and shortens the days and hours, so the index of the dial, entering by the points opposite the centre round which the wheel turns, by its daily motions, sometimes in greater, at other times in less periods, will pass through the limits of the months and days. The management of the water, and its equable flow, is thus regulated.

Inside, behind the face of the dial, a cistern is placed, into which the water is conveyed by a pipe. In its bottom is a hole, at whose side is fixed a brazen tympanum, with a hole in it, through which the water in the cistern may pass into it. Within this is inclosed a lesser tympanum attached to the greater, with male and female joints rounded, so that the lesser tympanum turning within the greater, similar to a stopple, fits closely, though it moves easily. Moreover, on the lip of the greater tympanum are three hundred and sixty-five points, at equal distances. On the circumference of the smaller tympanum a tongue is fixed, whose tip points to the marks. In this smaller tympanum a proportionable hole is made, through which the water passes into the tympanum, and serves the work.

On the lip of the large tympanum, which is fixed, are the figures of the celestial signs; above, is the figure of Cancer, and opposite to it, below, that of Capricornus. On the right of the spectator is Libra, on his left Aries. All the other signs are arranged in the spaces between these, as they are seen in the heavens.

Thus, when the sun is in the portion of the circle occupied by Capricornus, the tongue stands in that part of the larger tympanum where Capricornus is placed, touching a different point every day: and as it then vertically bears the great weight of the running water, this passes with great velocity through the hole into the vase, which, receiving it, and being soon filled, diminishes and contracts the lengths of the days and hours. When, by the diurnal revolution of the lesser tympanum, the tongue enters Aquarius, all the holes fall perpendicular, and the flow of water being thus lessened, it runs off more slowly; whence the vase receiving the water with less velocity, the length of the hours is increased.

Thus, going gradually through the points of Aquarius and Pisces, as soon as the hole of the small tympanum touches the eighth part of Aries, the water flows more gently, and produces the equinoctial hours. From Aries, through the spaces of Taurus and Gemini, advancing to the upper points where the Crab is placed, the hole or tympanum touching it at its eighth division, and arriving at the summit, the power is lessened; and hence running more slowly, its stay is lengthened, and the solstitial hours are thereby formed. When it descends from Cancer, and passes through Leo and Virgo, returning to the point of the eighth part of Libra, its stay is shortened by degrees, and the hours diminished, till, arriving at the same point of Libra, it again indicates the equinoctial hours.

The hole being lowered through the space of Scorpio and Sagittarius, in its revolution it returns to the eighth division of Capricornus, and, by the velocity of the water, the winter hours are produced.

(Vitruvius, *On Architecture* 9.8.8–15)

Physical remains of such complex machinery are rare. Fragments have been found of the star-plate of two such water clocks, both dating perhaps to the second century AD, one from Salzburg, which was monumental in scale at a diameter of over 1 m, and another much smaller one from Grand in Lorraine.[63] Inside the Tower of the Winds in Athens, built in the second century BC by Andronikos from Kyrrha in Macedonia, there are traces on the floor of channels for piping water from a reservoir set at the back of the building.[64] On the basis of Vitruvius' description of such water clocks, Stuart and Revett innovatively interpreted this piping as servicing a water clock. They pointed out how a suitable natural water source existed in the appropriately named stream Klepsydra, which ran from the north slope of the Akropolis, above the Tower, and then underground to resurface kilometres away at the coast near Phaleron (visitors to the Akropolis slopes can still walk down an alley named after this stream).[65] Judging from its mention by both Varro (*On Farming* 3.5.17) and Vitruvius (*On Architecture* 1.6.4–7), the 13 metre-high octagonal Tower was well known in antiquity. It was a *tour de force* of timekeeping instruments, incorporating not only a water clock inside but also nine sundials on its exterior walls and the reservoir annex. The influential reconstruction by Noble and Price of the interior water clock, which partially reflects Vitruvius' description of an anaphoric clock, is now disputed,[66] but modern wooden flooring currently makes alternative, detailed reconstructions difficult to propose.

In many of the cases noted above, we are witnessing a move towards standardised time. But this uniformity in the measurement of time remains limited in the ancient world, often staying within specific contexts, such as the legal or military worlds. The use of equinoctial hours on the plane sundial found at Oropos suggests an early attempt to establish standardised time in that particular site for some unknown purpose, but it seems not to have taken hold elsewhere, perhaps because of the complexity of producing the dials required in different places and the relative ease of producing alternative dial faces based on the celestial sphere.

A generation ago Price discounted the use of sundials and water clocks before the fourth century BC as timekeepers.[67] 'I take all this evidence not as that of time-determining practices, but rather as devices for reflecting the orderly sequences of the omen events – heliacal risings and settings, new moons, equinoxes and solstices.'[68] Timekeeping in his view – by which we need to understand 'telling the time' – was an innovation of the first century BC, which came in the train of new technological developments. Thus, he believed, the means came first, followed by the function.[69] The new means were part of the development of the old tradition of making models which duplicated natural phenomena. Sundials before the first centuries BC/AD had no hour numerals and sometimes not even hour lines, but did bear inscriptions identifying the placement of the solstices, equinoxes and zodiacal signs. So the sundial served primarily as a calendar, rather than as a clock, and in some forms, such as the spherical, they mirrored the apparent celestial dome in their very form. But from the first century AD, Price held, sundials became 'a timing device which is actually used for telling both the time of day and the time of year.'[70]

In the years since Price made this assessment, nothing, it would seem, has appeared to undo it significantly. He is still correct about the general lack of hour numerals on sundials before the Roman Imperial period. Gibbs counted only four sundials which indicate all the hours (by Greek letters), and she believed that they all probably date, like the sole, securely dated example from Ephesos, to the third century AD or later.[71] Eight further dials indicate by various marks the third, sixth and ninth hours, but none of these is firmly dated. In the literary record also it is from the first century AD that we find, in both Roman and Greek contexts, that the hours are numbered. Half hours were recognised from the fourth century BC, to judge from a fragment of Menander, but what instrument the poet had in mind and why he even cared to note them (was he telling the time?), we do not know. The use of very small fractions of the hour in *P. Hibeh 27*, dated ca. 300 BC, illustrates simply a theoretical, not practical, measurement among astronomers, and again the hours are not numbered. Aulus Gellius' quotation from Plautus in the second century BC does not number the hours. We shall find in the next chapter among writers from the Imperial Roman period, such as Martial, Pliny the Younger and Artemidoros, how characteristic it is then to recognise sharper definition in the subdivision of the day by means of the numbered hours.

But if we turn our attention to other evidence, in the mid-third century BC the Ptolemaic postal system operated with 'hour passes', which explicitly numbered the hours at which the courier reached the stations, specifically the hour before dawn and then the first, sixth, eleventh and twelfth hours.[72] What instrumentation was used to measure the hours is not detailed in the surviving logbook, but it does not matter whether it was a sundial or a water clock. The fact is that hours were numbered and counted: some people already in the early Hellenistic period were 'telling the time' in terms not very different from ours. Nor does it necessarily apply that even if individual hours were not numbered, then they were not counted. We do the same ourselves today with analogue clock or watch faces, which do not always number the hours, yet this does not prevent us from knowing the time of day as a function of the number of the hour.

In less time-pressured contexts, one could tell the time by quartering the day: by the end of the first third of the first century AD, the recording of events surrounding Jesus' death between the sixth and ninth hours (e.g. *Matthew* 27.45–50) is indicative of this popular practice of measuring time within the day at least in three-hour blocks.

The earliest planar dial and indeed the earliest surviving stone sundial of all – the equatorial dial from Oropos, dating to the second half of the fourth century BC – already has hour lines. The earliest surviving conical dial, that from Heraclea ad Latmum in Turkey, dating to the end of the third century BC, has hour lines.[73] And the earliest spherical dial, from Istros on the Black Sea, dating also to the third century BC, is a fragment which has the meridian and an hour line preserved.[74] It looks, therefore, as though hour lines came with sundial technology, and were well established in the third century BC, when in literature Kallimachos (fr.550 Pfeiffer) seems to be the first non-scientific writer to use the term *hōra* to signify 'hour' rather than 'season'.[75] There seems little point in marking out the hours on sundials if one is not going to use them to 'tell the time'. In that respect, then, Price's assertion that sundials before the Imperial Roman period were not true timekeepers, telling the time of day as well as the time of the year, looks over-refined when faced with the evidence.

What about time measurement? Edwards dated interest in the measurement of time in the day only from the period of Hipparkhos.[76] He argued that before the time of Hipparkhos and the introduction of equinoctial hours (which he dated to the mid-second century BC) the Greeks did not measure time within the day. The use of equinoctial hours standardises time so that we can measure time not only through the day (which seasonal hours permitted) but through the year on a uniform basis. The equinoctial hour is a measurement tool. But now with the Oropos sundial, equinoctial hours are to be dated to the latter half of the fourth century BC.[77] It seems unlikely that it is one thing to articulate the day into uniformly measured hours, but another to measure via these articulations.[78] So it appears that we should push the concept of measuring time back to the fourth century too.

Yet what would have given the impetus to such measurement then? The answer – on the large stage, as opposed to the small, esoteric stage of, say, magicians with their lamps – may lie in the contemporary development of means of measuring stretches of time in the political and military worlds, where equalisation of justice, on the one hand, and of responsibility on the other, stimulated the use of a standardised measure of time.[79] Despite its early appearance at Oropos, the equinoctial hour does not appear to have been more widely adopted until the second century BC, and then by astronomers for ease of calculation rather than in daily life. The seasonal hour will hold sway well into the medieval period, when the sand glass and the mechanical clock will cause the equal hour to supersede it. The prime cause for the adoption of the equinoctial hour in the medieval period was technological. We have seen, after all, that the main forms of stone sundial among the Greeks and Romans lent themselves to the persistent use of the seasonal hour. Once they are replaced by another mechanism (the mechanical clock), their measure of time was equally likely to be replaced too (by the equinoctial hour). Indeed, until the form of the clock changed, the ingenuity of antiquity was almost entirely diverted into finding clever ways of displaying the variable seasonal hours.[80]

The development of sundials and similar means of dividing the day and year into artificial segments can be paralleled by the development of calendars that achieve the same purpose and effects. Yet in both instances the abstraction from natural measures of time never creates a complete divorce. Underneath the complexities of built timepieces such as the Tower of the Winds in Hellenistic Athens, or of temporal constructs such as the Julian calendar in late Republican Rome, lie still visible vestiges of the natural cycles and phenomena from which they ultimately derived.

6
CONCEPTIONS OF TIME

With the principal surviving kinds of sundial – the spherical, conical and plane – it is still possible to discern the underlying cosmological structure which informs these dials. This is most obvious with the spherical sundial, which essentially represents the apparent dome of the sky in mirror image in its carved, concave surface (Figure 4.4). The relationship between dial and sky is initially less obvious with the conical (Figure 4.5) and plane (Figures 4.1, 4.3, 4.6) sundials, let alone the cylindrical, but between all of these the difference is fundamentally one of the form of projection of the celestial sphere on to variously shaped surfaces, rather than a radically different frame of reference. If we look more closely at the design principles of the spherical dial, we can discern the close relationship between this type and some of the other sundials. But we can also recognise an increasing abstraction from the observable world, which matches a growing conceptualisation of the cosmos as a geometrical entity.

Vitruvius (*On Architecture* 9.7.1–7) provides, through an *analemma*, a method for designing sundials, and despite the awkwardness and poor preservation of his text, the *analemma* is now well understood. He describes only the section, or elevation, of the sundial through the meridian. Particularly useful has been the modern addition to this of the method for laying out the seasonal and hour lines of the sundial, and the extension of the *analemma* to a variety of dial types.[1]

The *analemma* as presented by Vitruvius is a two-dimensional projection of the celestial sphere on to the plane of the meridian, and as such it betrays a close structural connection with celestial globes and armillary spheres.[2] It is likely that Eudoxos in the fourth century BC worked with a celestial globe, and certain that Hipparkhos did so by the second century BC.[3] No working model survives, so the closest we can get to imagining what one looked like in reality is through the artistic representations of celestial globes, notably the large-scale specimen borne by the Farnese Atlas, a Roman statue of the first or second century AD, but reflecting a Greek original of the second century BC.[4] This presents forty-one constellations in figural form, set against a backdrop of the parallel circles of the equator between the two tropics and the two polar circles (arctic and antarctic); the slanting ecliptic, with the broader zodiacal belt, runs from one tropic to the other across the equator; and from pole to pole run the equinoctial and solstitial colures across the equinoctial and solstitial points respectively on the ecliptic.

If we strip a celestial globe down to its bare essentials, and form just a framework of rings and hoops to represent the various circles of the equator, ecliptic, tropics and so on, leaving empty the spaces in between, we end up with the armillary sphere. This is simply a three-dimensional skeleton of the celestial globe. The earliest surviving image of one is in a mosaic floor panel from the so-called Casa di Leda in Solunto in northwest Sicily (Figure 6.1).[5] It dates probably to the late second century or early first century BC,[6] a little later than the period of Hipparkhos, and shows, as ribbon-like bands surrounding a spherical earth (a significant cosmological concept in itself), the equator and the two tropics crossed by the ecliptic, both polar circles, and the equinoctial colure.[7] If the outermost ring was originally graduated, as seems to have been the case, it may have served calendrical purposes, and permitted the actual instrument represented by this image to make calculations, for example, of day lengths for each day.[8] Clearly the representation of an astronomer's working instrument, it is an intriguing embellishment to a

Figure 6.1 Solunto, Casa di Leda: mosaic of armillary sphere.

Source: Drawing from von Boeselager 1983: Taf. XV Abb. 29; reproduced by kind permission of G. Bretschneider Editore.

private house's floor decoration, on an island where one of the outstanding astronomers of antiquity, Arkhimedes, had chosen to live (and die) a century earlier.[9]

The three-dimensional physical model on which the mosaic image is based obviously must be of earlier date. There are hints of a construction of this type as early as Plato in the fourth century BC, when he describes, in his *Timaeus*, the creation of the world by the creator Demiurge, who

> separating the whole mixture into two down its length, and laying each with the other like *khi*, centre over centre, bent them into a circle, joining them together, to themselves and to the other at a point opposite the join. And he encompassed them in uniform motion revolving about the same axis, and made one the outer and the other the inner circle.
>
> (Plato, *Timaeus* 36b6–c4)

A band of created material is split lengthways into two strips, which are then bent round to form two circles. These are then placed one across the other, like the Greek letter *khi* (X), creating the celestial equator and the ecliptic.[10]

The step from a three-dimensional model of an armillary sphere to a two-dimensional representation of it was a relatively short one in an age when perspective and particularly foreshortening were understood. The step from the two-dimensional image to a projection of its parts unfolded, which is fundamentally what Vitruvius' *analemma* is, is also a small one in terms of the conceptual abstraction required. It is, however, a considerably larger leap to understand the complexities of plotting the seasonal and hour lines which must cross the meridian plane on various types of sundial. Vitruvius (*On Architecture* 9.7.7) says that he has avoided adding this extra step 'not from laziness, but to avoid offence by writing too much'. This is a pity for us, but subsequent manuscript copyists no doubt were grateful.

The basic steps in the design of both the spherical and conical types, and their interconnectedness, are demonstrated in the following *analemma*, whose argument is based on that of Vitruvius (*On Architecture* 9.7)[11] (See Figure 6.2). To explain:

1. Draw a line on a plane surface. Mark on it a point A. Call this line the horizon.
2. Draw out a circle, centred on A, with a radius of 9 units.
3. Mark out one radius along the horizon. Call the end point I.
4. Drop another radius below and perpendicular to the horizon. Call the point where it intersects the circle B.
5. Draw a line perpendicular to AB (and so parallel to AI). Mark out 8 units along this line. Call the end point E′.

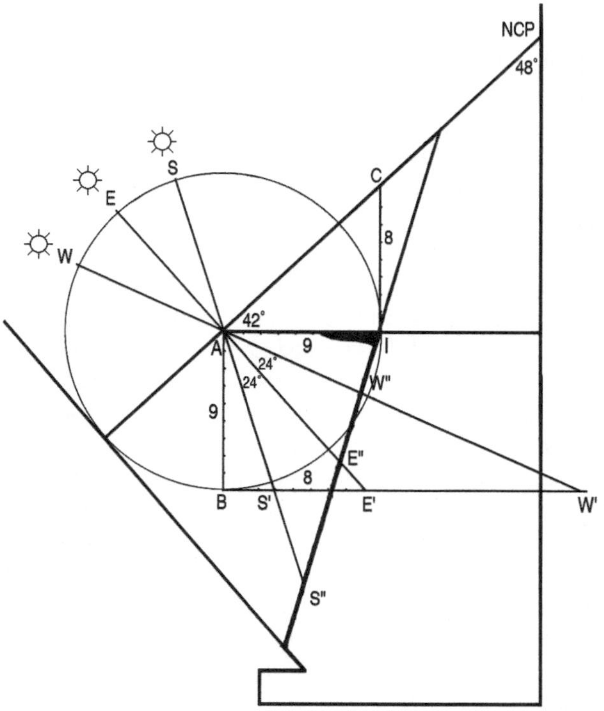

Figure 6.2 An *analemma* for the section of a spherical and a conical sundial for the latitude of Rome.

Source: Drawing R. Hannah.

6 Draw a line from E′ to A, and extend it outwards to E.

7 Similarly, draw a line up from I, perpendicular to AI (and so parallel to AB). Mark out 8 units along this line. Call the end point C.

8 Join AC and extend this line outwards. This is the *axōn*.

9 Another approach, once E′AE has been drawn in step 6, is to extend a line perpendicular to E′AE from A, thus creating the *axōn*, and then to draw a line from I perpendicular to AI, to intersect the *axōn* at C. It will be found that IC is 8 units long. The angles created, BÂE′ and IÂC, are also by definition the same, belonging to equal triangles.

10 Vitruvius (*De arch.* 9.7.1) says that at the time of the equinoxes in Rome, a *gnōmōn* of 9 units will cast a shadow of 8 units. This is the principle underlying the construction of triangle ABE′, in which AB would be the *gnōmōn* of 9 units on the plane sundial, and BE′ the noontime shadow of 8 units. This principle may be based on empirical evidence, but the theoretical and more general underpinning is easier to see from triangle AIC, which is equal to triangle ABE′. In triangle AIC, AI is the horizon.

Angle IÂC is generated by the sides of 8 and 9 units, and is 41.6° in size, by measurement, or by trigonometry: $\tan^{-1}(8/9) = 41.6°$. By geometry, a line that points at the celestial pole makes an angle with the horizon that is equal to the latitude of the place where the line is installed. The actual latitude of Rome is 41.9°. So AC must point in the direction of the North Celestial Pole, and the angles IÂC and BÂE′ represent (in practical terms) the latitude of Rome. By the same token, line E′AE, being perpendicular to AC, must represent the celestial equator, on which the equinoctial points sit.

11 To either side of AE′ mark out an angle of 24°, to represent the angular separation of the ecliptic at the solstices. Draw the lines of these angles to intersect BE′ at S′ and BE′-extended at W′. These represent the shadows cast at noon at the times of the summer (S′) and winter (W′) solstices.

12 Arc BI represents the noontime line or meridian of a simple spherical sundial. It is also now crossed by the solstitial and equinoctial lines. The two arcs thus created on either side of the equinoctial point are equal.

13 Line AC, the *axōn*, may be treated as the axis of a cone. Its base will be constructed parallel to the equatorial/equinoctial line EAE′. In the diagram, I have arbitrarily set the base simply as a tangent to the circle, and then given an angle of 60° to the side of the cone.[12]

14 AW′ intersects the side of the cone at W″, and AE′ at E″, while AS′ is extended to intersect at S″. W″E″ is therefore shorter than E″S″.[13]

15 The equinoctial arc which will run through point E″ on the dial is circular, since it is formed by a section through a right cone that is parallel to the base and perpendicular to the axis. But the solstitial arcs will be elliptical, since they are formed by sections that are angled more or less than perpendicular to the axis.

16 It would be possible to construct the cone so that its side is perpendicular to the line AW′, thus making the winter solstice arc parabolic, but other than making the end-product seem more elegant geometrically, I see no constructional advantage in this.

It will be clear from this hypothetical *analemma* that the conical sundial is intimately related by geometry to the spherical type, and represents a simplification in construction terms of the latter.

The proliferation of sundials in a wide variety of geometrical forms from the late Classical period onwards – over 340 sundials of different kinds are now known[14] – is indicative of an increasing abstraction in the way time was measured and perceived. Whether from patterns on the ground based on the human shadow, or from mirror-images of the celestial dome designed to recapture the apparent movement of the sun across the sky, the development of Greek and Roman sundials followed a route which distanced their users further and further from the natural means of time measurement. The denaturisation which we witnessed in the skits of Greek and Roman satirists is matched in the very forms of the instruments which were used to mark the passage of time. It becomes harder and harder to see the natural or the human in the end-products. Yet, as Cassiodorus indicated (*Letters* 1.46.2), this distancing from nature comes to be seen as a mark of civilisation, something that distinguishes humans from animals. Sundials carry an extraordinary evolutionary responsibility.

The degree of abstraction which was quickly achieved by Greek diallers is well illustrated in the contrast between the two broadly contemporary sundials found at Ai Khanoum, the Greek city founded at the extreme east of the Hellenistic world in Afghanistan. These were certainly in use in the mid-second century BC, in the last phase of the city before its destruction, but they may have been manufactured earlier, in the third century.[15] The spherical dial has already been mentioned because of its network of lines dividing the day into twelve hours and the year into twelve solar, zodiacal months. The concave form of the dial still recalls the apparent domical form of the sky above, while the inscribed lines track the observable path of the sun through the day and the year via the shadow of the *gnōmōn*. Nature is not very distant from the mind of the dialler here. With the cylindrical sundial from the same site, on the other hand, we encounter a form which is unique so far in the literary and archaeological records, and which is clearly a considerable conceptual distance from the natural world. This sundial consists of a block of marble, taller than it is wide, out of the centre of which a cylindrical hole has been carved. The inner surface of this hole has been graduated with two sets of straight lines, a set emanating from each of the broad faces of the slab and radiating towards the interior of the hole. The stone is of unequal length on its front and back faces, and bevelled at its base, so that it did not stand upright, but was set at an angle. Measured from the vertical, that angle is 37°4′, which is practically the latitude of Ai Khanoum (37°10′). With the slab's longer face set towards the north and its shorter to the south, the angular fix causes the stone to be parallel to the celestial equator. This means that the sun would have shone into the southern aperture of the cylindrical hole from the autumn equinox, through the winter solstice, and on to the spring equinox, and then into the northern interior of the hole from the spring equinox, through the summer solstice, and on to the autumn equinox. Because of its being set parallel to the celestial equator, the stone would have borne no shadow at all on the days of the two equinoxes themselves. The lines inscribed within the hole then told the time of day in hours, but the season of the year only with regard to the solstices and equinoxes, which marked the inner and outer extremities respectively of the interior lines. In that sense, it was a less precise calendar than its spherical cousin, even though it was a more complex form of dial.

The idea of creating a sundial by setting a stone slab up at an angle relative to the latitude of the locality (taken either as the latitude itself if measured from the vertical, or as the co-latitude if measured from the horizontal) had already been seized upon in the second half of the fourth century BC on the Greek mainland, when the equatorial dial from Oropos in Attika was made.[16] This sundial slab stood at an angle to the horizontal of 50° (Figure 6.3). The complement of this is 40°, which is very close to the latitude of Oropos (38° 19′), suggesting that the dial was made expressly for the site, or more specifically for the sanctuary of the Amphiareion at Oropos, where it was found.

In this case, though, the slab was not punctured by a hole, but had the upper halves of its northern and southern faces inscribed with a large semicircle on the northern face, and a small and a large concentric semicircle on the southern side, which also bore an explanatory and dedicatory inscription. This inscription, unusually, followed the circuit of the outer semicircle. It is restored as follows by Schaldach on the basis of Petrakos' original reading (with my literal translation to show the restorations):

> When on the] circle near the *gnōmōn* the shadow [goes, in this part of the sundial it indicates the winter | solstice;] when across the biggest circles the shadows [go, they indicate the equinoxes; this side | indicates] the shadow of the *gnōmōn*, when the autumnal [equinox occurs, whereas when the spring equinox occurs, the shadow disappears.]

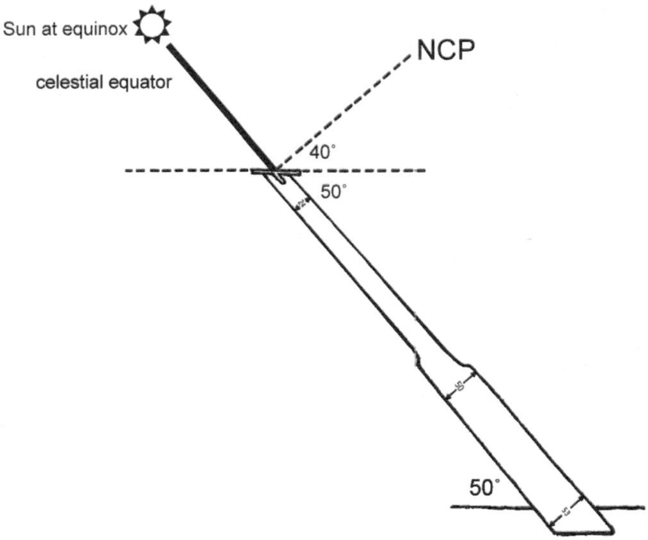

Figure 6.3 Piraeus, Archaeological Museum, inv. 235: plane sundial from Oropos, showing the sun at the equinoxes and demonstrating that its intended latitude is 40°.

Source: After a drawing from Schaldach 2006: 112; reproduced by kind permission of Verlag Harri Deutsch.

Theophilos	Th . . .
[erasure]	[erasure]
Athenian	G[. . .]

If the restoration is correct, or at least on the right tracks, it indicates what can be inferred anyway from observation of the circles when the slab is placed at the bevelled angle in a stand or in the ground: the small semicircle on the side bearing the inscription represents the *gnōmōn*'s shadow at the winter solstice, and the two large semicircles indicate the shadows at the equinoxes. In fact these outer semicircular arcs only approximated the equinoxes – because the slab stood parallel to the celestial equator, at the equinoxes the *gnōmōn*'s shadow would have disappeared completely had not the stone been left thicker at those points so as to catch it. On what must be the southern face, which bears the inscription, the small semicircle marked the winter solstice; a corresponding small semicircle, which would be expected to mark the summer solstice on the northern face, is oddly absent. Figures 6.4 and 6.5 illustrate the geometry underlying the theoretical placement of the solstitial lines. The semicircles on each side were subdivided by a network of radiating lines, marking the hours of the day. The northern face's radiating lines also appear 'unsteady', as if the mason had incomplete directions for this face of the dial.[17]

This dial from Oropos has been used by Schaldach to refute the traditional assumption that spherical sundials were the earliest type to be developed in

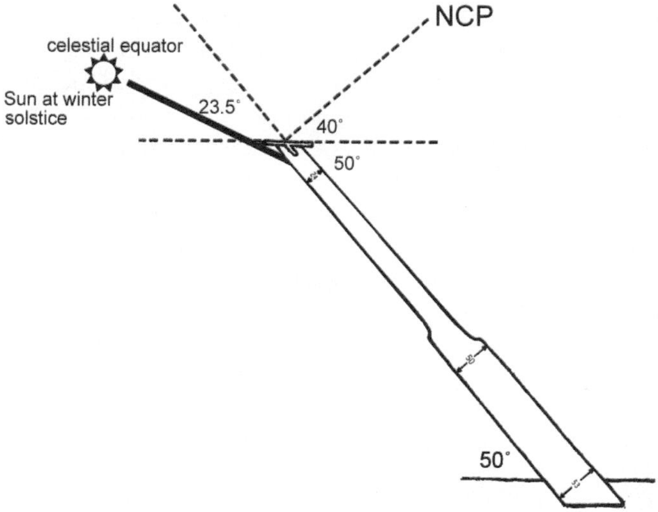

Figure 6.4 Piraeus, Archaeological Museum, inv. 235: plane sundial from Oropos, showing the sun at the time of the winter solstice.

Source: After a drawing from Schaldach 2006: 112; reproduced by kind permission of Verlag Harri Deutsch.

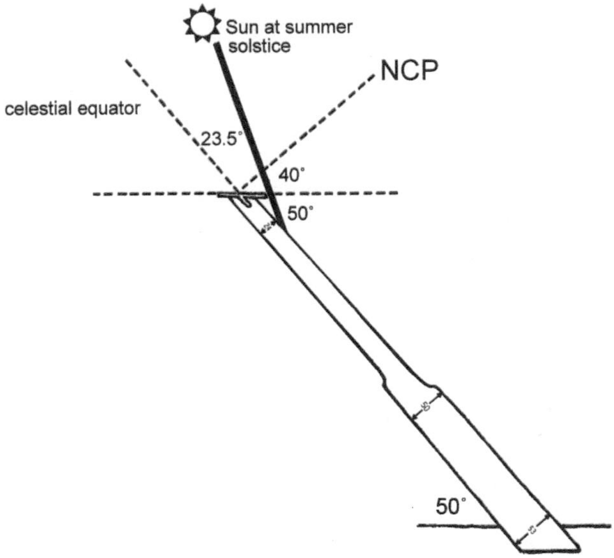

Figure 6.5 Piraeus, Archaeological Museum, inv. 235: plane sundial from Oropos, showing the sun at the time of the summer solstice.

Source: After a drawing from Schaldach 2006: 112; reproduced by kind permission of Verlag Harri Deutsch.

Greece.[18] We also have already had occasion to question this belief. Instead, we may now think in terms of plane dials with engraved semicircles as the ancestral form for Greek sundials in general. Precursors of this form had long existed in the Near East and Egypt, and may have provided the model for the Greeks.[19] The step from making such dials to lie flat on the ground or to rest vertically on walls, to fixing them on a base so that they were angled in alignment with the local latitude is a significant conceptual and technical achievement, about whose evolution we know practically nothing. It must have happened, one would imagine, by the first half of the fourth century BC: awareness of the ecliptic (the apparent path of the sun, moon and planets) had dawned on the Greeks by the end of the fifth century, to judge from the names of astronomers to whom the discovery was attributed, so there was probably present by then a geometrical conceptualisation of the cosmos, which would seem to be a prerequisite for the further step of realising the theory which underpins a sundial like that from Oropos.[20] One wonders about the role in this development of the astronomer, Eudoxos, who happens to be the earliest inventor of sundials in Vitruvius' list.

A further curiosity with regard to the Oropos dial is that the hour lines suggest that they do not represent the normal 'seasonal' hours of antiquity, which are the result of dividing daylight by 12 throughout the year, and are therefore unequal from one day to the next. Instead they show equinoctial hours, which remain the same length throughout the year.[21] The invention of equinoctial hours must belong to the fourth century BC and not later, more or less in the time of Eudoxos.[22] Their use in the *Hibeh Papyrus* (*P. Hibeh* 27), where they are subdivided down to 1/45th of an hour, implies an earlier date than ca. 300 BC and demonstrates an astronomical function.[23] So we know that astronomers used such equal hours in their calculations, but we would not expect to find them in a semi-public, but still 'non-scientific', context such as the religious sanctuary of the Amphiareion, nor is it clear why they were preferred there.

Just how unscientific the context at Oropos was is demonstrated by the sundial's inscription. This presents a simple set of instructions, a basic 'user's manual', to explain how to 'read' the dial, rather than providing any description of why the dial works in the way it does. In other words, it is not a teaching tool, such as an astronomer might use in training a student, nor are its diurnal or annual lines sufficiently precise to measure more than the hour and the solar tropics. Rather, it is a practical, utilitarian device for the ordinary, albeit literate, visitor to the sanctuary to use. How it was used we can only guess. A festival was held every four years at the sanctuary in honour of the healer-hero Amphiareus, comprising athletic and equestrian contests, and a theatre hosted dramatic performances of some kind.[24] Sundials were also associated with the Theatre of Dionysos in Athens, where a very large conical one can still be seen above the seating (Figure 0.1).[25] In this latter case it seems to me perfectly possible for spectators to have used the sundial to time performances by the hour through the day, so clear are the hour lines on the dial face even from the front rows near the orchestra, where the officials sat. The Oropos dial, on the other hand, is small, and needs to be read from close by, nor is it clear what association it had with any of the festival activities at the sanctuary. Nevertheless, both locations demonstrate the popularisation of timekeeping technology from the early Hellenistic period.

Sundials helped embed into people's consciousness the concept of the seasonal hour, but they did little, it seems, to encourage or enable people to think in smaller portions of time. The rare 'half-hour' first appears in surviving literature in a fourth-century BC comedy by Menander, to judge from a stray surviving fragment (fr. 1015). This fraction of the hour is inscribed on only a couple of surviving Greek sundials, suggesting such refinements in time really were of little interest either in private or public contexts.[26] In his instructions for making sundials in the third century AD, Cetius Faventinus (29.2) remarked that almost everyone is 'in too much of a hurry to ask more than what hour it is'.[27] Calculations derived from some sundials, which were carried a considerable distance from one latitude to another and yet still used, have suggested that people would tolerate a quarter of a seasonal hour's inaccuracy in the telling of local time.[28] If this is so, then ancient expectations of accuracy and punctuality (and indeed other premodern and some modern, indigenous expectations) differ markedly from those in modern, industrialised societies. In the Roman world, which displays greater interest in timetabling than the Greek world generally did, none of the surviving advertisements of gladiatorial combats provides a starting time – spectators just showed up whenever.[29] Even in scientific contexts, inadequate means of measuring small amounts of time leaves Hero of Alexandria (*Dioptra* 4) with only a sundial to calculate the rate of flow of water, by letting water flow into a large basin for an hour.[30] For more rapid movement, there was simply no mechanism for Hero to measure the speed of revolution of his newly invented steam machine. Later, Frontinus measures the flow of water through an aqueduct simply by the capacity of the pipe, not by any timed rate.[31]

What was lacking in technology for small measures of time was also lacking in vocabulary. Roman writers referring to small periods of time had recourse to awkward phrases like 'a finger of time' (*articulus temporis*: Cicero, *For Quinctius* 19 – the word *articulus* usually refers to small joints in the body). This, however, may be nothing more refined than half an hour (*semihorae articulum*: Cicero, *For Rabirius* 6). Pliny (*Natural History* 2.58) refers to a twenty-fourth of an hour as *semuncia horae*, using the terminology of coinage (where the *semuncia* was a twenty-fourth of an *as*) to do duty for what was lacking with reference to time. On this basis, Houston argued that this absence of a specialised vocabulary signified quite simply a lack of awareness of small periods of time in the Roman mentality.[32]

Nonetheless, we should not be surprised that the concept of the half-hour was well embedded by the first century AD in the consciousness of astronomers and others in the 'scientific' community, such as geographers and doctors, who often specify that it is the equinoctial half-hour that they are measuring. But it is interesting to see that it had also penetrated sufficiently into the popular mind as to find its way into Christian apocalyptic literature at the same time: 'And when he had opened the seventh seal, there was silence in heaven, as it were for half an hour.' (*Rev.* 8:1) At that time, according to a long tradition, an equinoctial half-hour, literally, may be meant.[33]

Throughout the Greek and Roman worlds, the popularity of sundials extended to both public and private contexts: not only religious and civic centres afforded access to these ancient clocks, but private individuals also had them at home. Their ubiquity is well captured in a quip, attributed to the Roman emperor Trajan early in the second century AD, which makes a real dial of the human face: 'If you put your nose facing the sun and open your mouth wide, you'll show all the passersby the time of day.' (*Palatine Anthology* 11.418)

One of the earliest surviving of all Greek sundials, after that from Oropos, comes from a private house in Delos and dates to the third century BC.[34] The second-century Tower of the Winds in Athens, with each of its eight faces and an annex decorated with vertical sundials, survives still in its original, public situation near the later Roman Agora.[35] On architectural grounds, eight of the external sundials, one on each outside wall, are now accepted as an original element of the design, despite the absence of any mention of them by Varro (*On Farming* 3.5.17) or Vitruvius (*On Architecture* 1.6.4–7) – an omission that once led to the suspicion that they may have been added afterward.[36] Also surviving *in situ* is one of the largest conical dials, long known in the archaeological record since its publication by Stuart and Revett in the late eighteenth century, above the Theatre of Dionysos on the south slope of the Akropolis in Athens (Figure 0.1).[37] The association of timing devices with a performance-centre is encountered also at the Amphiareion at Oropos, where we have already met the equatorial dial, and where a water clock was built in the fourth century BC, perhaps to check the accuracy of the sundial and, together with the sundial, to assist in timing rituals or other performances at the theatre or during the games at the four-yearly festival.[38]

It is well to remember that the geometrical complexity of some sundials, both large and small, is more than matched by a continuing production of simple dials, or by plain ignorance, or even by the supersedence of a different way of conceiving time. Even if there ever was any evolutionary development in the forms of sundials, there still remained a demand for the simpler types. The complexity of dials like those from Ai Khanoum and Oropos can also mask the incomprehension that some diallists must have felt when faced with the geometrical theory required for some of their products. The sophisticated and otherwise expertly carved Oropos dial has suggested this, with its less-than-ideal, northern, 'summer' face. It is plain that for other diallists theory greatly outstripped their understanding: the mathematics demanded by a given form was not always matched by the execution in the end-product. This is particularly the case in the Roman period.[39]

The lack of care or comprehension is well illustrated by a well-known, conical sundial from Alexandria now in the British Museum. This dial was found in 1852 at the foot of the obelisk now known as Cleopatra's Needle; both were transported to London.[40] The original placement of the *gnōmōn* on the sundial is clear, even though it has disappeared, because of the remains of its hole: it was set horizontally over the face of the dial, and its original length can, in theory, be determined either graphically or mathematically from the seasonal day-curves on the dial face.[41] Three curved, seasonal, lines are engraved on the face's surface, and both literary instructions and inscribed surviving examples would suggest that these should represent the solstitial and equinoctial lines for the latitude of the site where this dial was intended to work.[42] But in reality on this dial they do not match where we would place the solstitial and equinoctial lines for the latitude of Alexandria, which other measurements indicate is where the sundial was intended to operate.[43] As it is, if the sundial stood in Alexandria and the shadow cast by the *gnōmōn* struck the presumed equinoctial line at the time of the equinoxes, then at the time of the winter solstice the shadow would never have reached the upper ('winter solstice') line, and it would have overshot the dial completely beyond the lower ('summer solstice') line at the time of the summer solstice. While the lower line just might signify a day in the year other than the summer solstice, such as a day just before or after that date,[44] any confidence that this may be so is undermined by the fact that the upper line can serve absolutely no calendrical role, since the *gnōmōn*'s shadow could not reach it at this latitude. In fact, one gets the impression that one or both 'solstitial' lines were cut more as guidelines for carving the separate hour lines, since the upper, 'winter solstice' curve is dotted with evenly spaced holes which mark the top ends of the hour lines.[45] So it appears that the dial was intended for Alexandria, yet of its engraved seasonal lines only the equinoctial serves its purpose accurately.

The issue can be represented graphically, taking a view through the plane of the meridian (See Figure 6.6). Here the outline indicates the profile of this conical dial along the meridian, and the three marks along the front face on the left indicate the sites of the three day-curves along the meridian line.

In Figure 6.7, using a template right-angle and assuming the middle day-curve (E) is the equinoctial line, one can establish graphically the theoretical extent of the *gnōmōn* along the extension of the top face of the sundial (to the point G), and, perpendicular to the equinoctial line, the direction of the North Celestial Pole (NCP). The angle thus created above the horizontal to the pole is 31°, which is the latitude of Alexandria, thus confirming the day-curve line at E as the equinoctial line. Two further lines (GW and GS), 24°

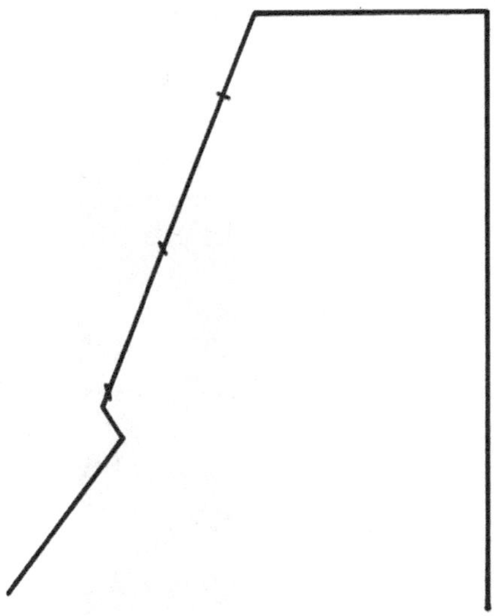

Figure 6.6 London, British Museum 1936.3–9.1: conical sundial from Alexandria; profile through the meridian indicating the points of the three day curves.

Source: Drawing by R. Hannah.

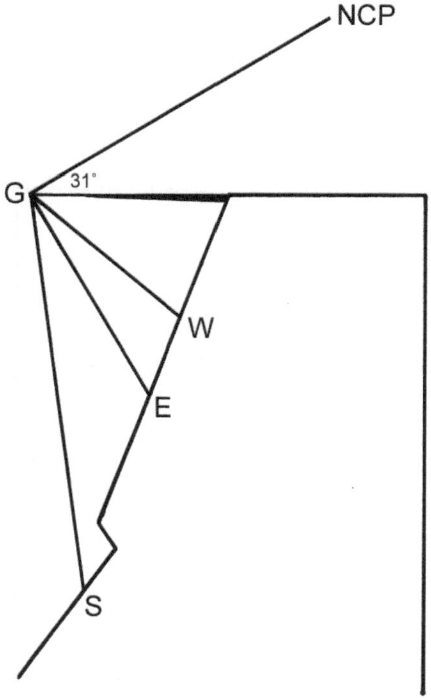

Figure 6.7 London, British Museum 1936.3–9.1: conical sundial from Alexandria; profile through the meridian, with graphically derived reconstruction of the *gnōmōn* to point G. Taking GE as the line of the equinoctial shadow, a perpendicular line from it shows the direction of the North Celestial Pole, at an angle of 31° to the horizontal, and delivers the intended latitude of the sundial, i.e. Alexandria. Lines GW and GS indicate where the winter solstice (W) and summer solstice (S) day curve lines should be inscribed on the meridian for the latitude of Alexandria.

Source: Drawing by R. Hannah.

on either side of the equinoctial line, indicate where the winter solstice (W) and summer solstice (S) day-curve lines should be inscribed for the latitude of Alexandria, if the equinoctial line is in the correct place.

In Figure 6.8 the lines GW_1, GE and GS_1 indicate the actual day-curve lines projected to G, the theoretical end of the *gnōmōn* of the dial. These day-curves make no sense as solstitial lines since their angular distances from the equinoctial line (at E) are not 24° (Vitruvius' 1/15th of a circle). Instead, angle $E\hat{G}W_1$ is about 36°, and angle $E\hat{G}S_1$ is about 19°.

Figure 6.9 shows Figures 6.7 and 6.8 overlaid, in order to demonstrate the differences.

In Figure 6.10, G_1 indicates the theoretical terminus (achieved graphically) of the *gnōmōn* for this sundial's actual solstitial lines (G_1W_1 and G_1S_1). But

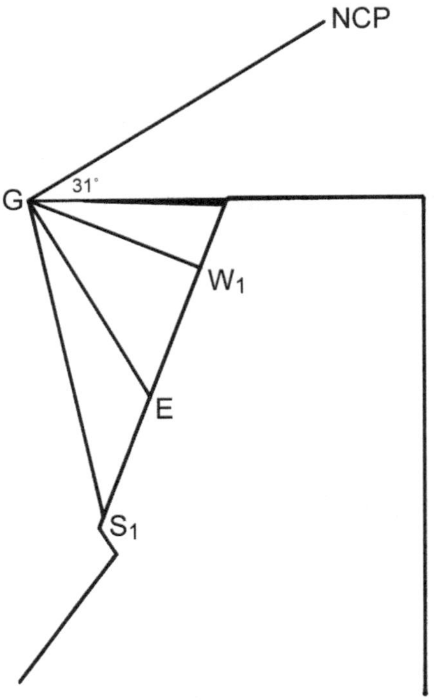

Figure 6.8 London, British Museum 1936.3–9.1: conical sundial from Alexandria; profile through the meridian. The lines GW_1, GE and GS_1 indicate the actual day curve lines projected to G, the theoretical end of the *gnōmōn* of the dial. Angle $E\hat{G}W_1$ is about 36°, and angle $E\hat{G}S_1$ is about 19°.

Source: Drawing by R. Hannah.

this reconstruction creates a different equinoctial line (G_1E_1: angle $E_1\hat{G}_1W_1$ = angle $E_1\hat{G}_1S_1$ = 24°). This in turn produces a different theoretical latitude of 51°N.[46]

The intended latitude of the sundial can be cross-checked against the angle of the face of the dial.[47] The required angles for the calculation are γ and σ in the drawing (Figure 6.11), and these can be measured:

In Figure 6.11, by geometry of triangles:

$$\sigma = \varphi + \omega, \text{ and } \omega = 90° - \gamma;$$

therefore,

$$\varphi = \sigma - \omega = \sigma - (90° - \gamma) = \sigma + \gamma - 90°;$$

by measurement,

$$\sigma = 69°, \gamma = 52°;$$

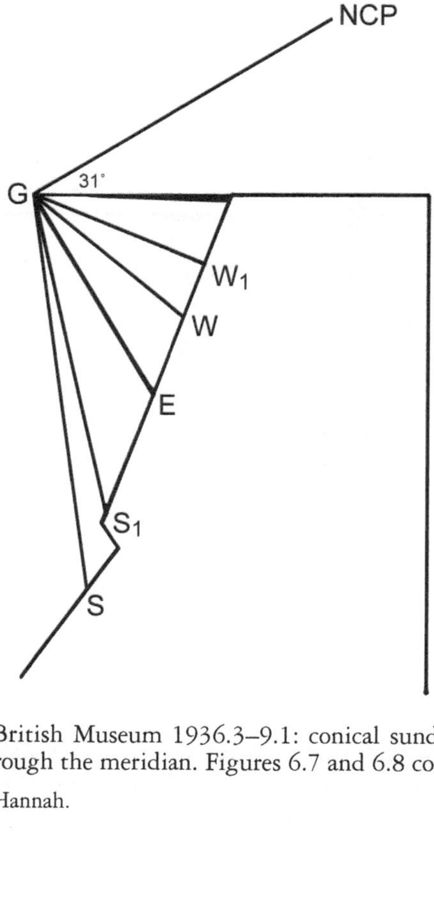

Figure 6.9 London, British Museum 1936.3–9.1: conical sundial from Alexandria; profile through the meridian. Figures 6.7 and 6.8 combined.

Source: Drawing by R. Hannah.

therefore,

$$\varphi = 69° + 52° - 90° = 121° - 90° = 31°,$$

which is the latitude of Alexandria.[48] So it appears that the dial was intended for Alexandria, yet of its engraved seasonal lines only the equinoctial serves its purpose accurately.

The association of the sundial from Alexandria with an obelisk is suggestive: the former tells the time from the sun, while the latter was a recognised symbol of the sun (Pliny, *Natural History* 36.64). The curiosity is that, as far as we can tell, obelisks were not used by the Egyptians as parts of sundials themselves,[49] however useful others found them for similar purposes – the emperor Augustus, for instance, made outstanding use of one in 9 BC on the Campus Martius in Rome (Pliny, *Natural History* 36.72), as the *gnōmōn* for the huge meridian line set up to commemorate both the takeover of Egypt in

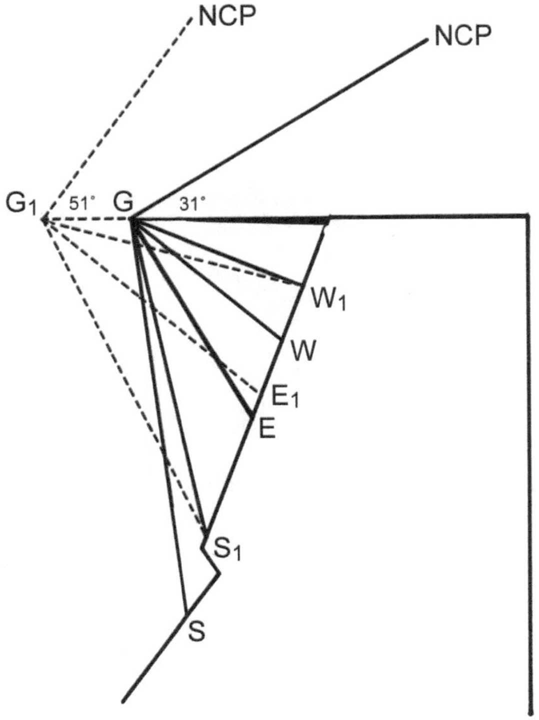

Figure 6.10 London, British Museum 1936.3–9.1: conical sundial from Alexandria; profile through the meridian. G_1 indicates the graphically-derived terminus of the *gnōmōn* for this sundial's actual solstitial lines (G_1W_1 and G_1S_1). But this reconstruction creates a different equinoctial line, G_1E_1, which produces a different theoretical latitude of 51°N.

Source: Drawing by R. Hannah.

30 BC and the correction of the Julian calendar;[50] but the designer, Novius Facundus, to judge by his name, was not an Egyptian.

Indeed, it was the Greeks and Romans who were adept at making time-keepers do double duty on both practical and metaphysical levels. Both of the sundials at Ai Khanoum were discovered in the gymnasium. There they would have been used to assist in telling the time for various activities, not just what we would term 'gymnastic' for physical exercise, but including teaching astronomy, a core subject in ancient education, as gymnasia in antiquity were much broader educational facilities than their modern equivalent.[51] Lucian's much later description of a Roman baths building includes 'two ways of showing the time, one through water and noise, the other telling it through the sun' (Lucian, *Hippias* 8), presumably referring to the later type of hydraulic water clock as well as a sundial.[52] Stuart and Revett long ago noted the parallelism between this instance of two types of clock

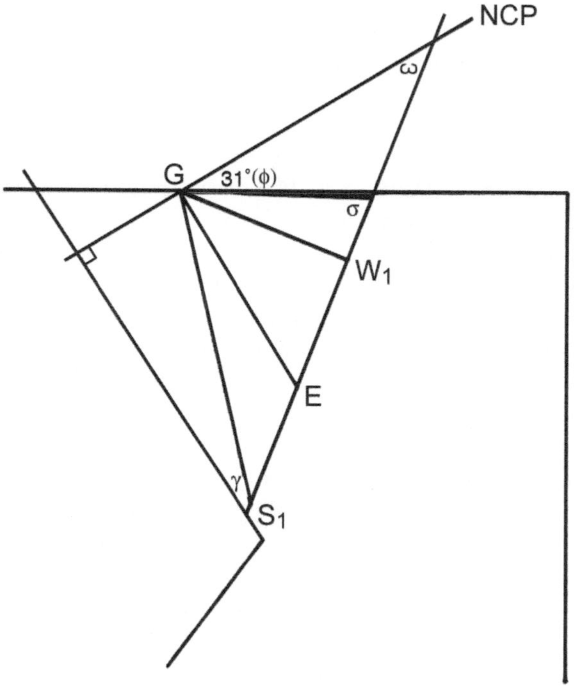

Figure 6.11 London, British Museum 1936.3–9.1: conical sundial from Alexandria; profile through the meridian. Calculation of intended latitude by measurement and trigonometry.

Source: Drawing by R. Hannah.

juxtaposed and the incorporation of a water clock and sundials in the single building of the Tower of the Winds near the Roman Agora in Athens.[53]

In these cases, the sundials still served primarily secular functions, but in other instances we find dials closely associated with religious sanctuaries. The dial from the Amphiareion at Oropos has already figured in our discussion. We may set this instance beside a large sundial at Klaros in western Turkey, which was set up beside the Temple of Apollo by a public official, the *agoranomos*.[54] In Delphi inscriptions indicate that sundials were also set up on columns at that sanctuary of Apollo.[55] There was a similar cultic collocation of a sundial on a column, donated by two magistrates perhaps in the time of Augustus, in front of the Temple of Apollo near the main Forum in Pompeii.[56] What appears to be happening at these sites is an actualisation, through the cultic furniture, of the identification of Apollo with Helios the Sun god, an identification which developed from the late fifth century BC onwards.[57] Within that context, it is then deemed appropriate to provide a timekeeping instrument which is intimately connected with the sun. These dedications also demonstrate clearly that the desire for civic philanthropy ('euergetism'), which was so strong a feature of both Greek and Roman life, could reasonably find expression in the setting up of a public timepiece. Such benefactions were worth recording even in death: from the first-century AD comes a private tombstone at Talloires in southern France, which commemorates the deceased's donation of a public *horologium* (probably a water clock) and an attendant slave to look after it, plus its own building and decorations, all at great personal expense.[58] The practice of such donation has survived well beyond antiquity.[59]

The excavations in Pompeii have provided us with thirty-five sundials. In the Greek east, the small island city of Delos has given a further twenty-five dials. Between them, the two sites demonstrate the popularity of the instrument from the Hellenistic period into the Roman Empire.[60] If, as we have had occasion to notice, Roman comic playwrights in the early second century BC could raise a laugh at the notion of a town being 'stuffed with sundials' which controlled ordinary life (Aulus Gellius 3.3.5), then, even if the joke were a transplant from Greek New Comedy (which is unlikely), it is probable that Rome itself was already in the middle Republic home to increasing numbers of sundials.[61]

This suspicion is confirmed by what we know of the earliest sundials in Rome, where they appear from the start of the third century BC. The first dial came to Rome in uncertain circumstances, as Pliny the Elder tells us:

> The first sun-dial is reported by Fabius Vestalis to have been erected eleven years before the war with Pyrrhus, at the temple of Quirinus by L. Papirius Cursor, when he dedicated it after it had been vowed by his father. But he does not indicate the principle of the dial's construction or the artist, nor where it was brought from or in whose writings he found this.
>
> (Pliny, *Natural History* 7.213)

The date of installation should be 293 BC,[62] but whether the sundial was intended to be used at all is not clear. Its dedication at the temple of Quirinus by Papirius at this time is highly suggestive of war booty,[63] which is precisely what the next sundial listed by Pliny was. This was brought to Rome from the sack of the Greek Sicilian city of Catania in 263 BC. Pliny tells of the Romans living for ninety-nine years in ignorance of the inaccuracy of this sundial for the latitude of Rome:

> M. Varro relates that the first public sundial was set up on a column behind the Rostra in the first Punic war by the consul M. Valerius Messala. It was carried off after the capture of Catania in Sicily, thirty years after the events relating to the dial of Papirius, and in the year of Rome 491. Its lines did not agree with the *horae*, yet they obeyed

> it for ninety-nine years, until Q. Marcius Philippus, who was censor with L. Paulus, put nearby one which was more carefully regulated; and this gift during his censorship was most gratefully accepted.
>
> (Pliny, *Natural History* 7.214)

It is worth noting that this sundial was set up for public use, in a public area rather than a sanctuary, as early as the mid-third century BC. It has recently been thought characteristic specifically of the Romans to desacralise and democratise timekeeping in this way, providing everyone with the capacity to tell the time.[64] But as the sundials from the gymnasium in Ai Khanoum demonstrate, a similar tendency was also displayed by the Greeks.

This anecdote from Pliny depends for its force on the dial's engraved lines not agreeing with the *horae* appropriate for Rome. Yet, ironically, while modern commentators have emphasised the four-degree difference in latitude between the two cities and have persisted in translating Pliny's *horae* as 'hours', the difference in latitude affects the hours of the day far less than it does the days of the year. Gibbs has calculated that for a certain popular type of sundial in the summer there would be a negligible error of just 4.2 minutes (0.07 hours) in the day, whereas the *gnōmōn*'s shadow in Rome would never have fallen on the Catania dial's summer solstice line, and would have fallen on the winter solstice line twice.[65] But while the very minor discrepancy in the daily hours has been noticed, no-one seems to have realised that Pliny may not have intended this to be the point of the story. An error of just a few minutes in the day would not have been noticeable on an ancient dial unless it was of enormous size. Rather, it seems likely that it was really the time of year, and not the hour of the day, that was recognised by the Romans as being inaccurately measured by the sundial. To achieve this interpretation all we have to do is alter the traditional translation of Pliny's *horae* as 'hours' to the less common but alternate meaning 'seasons' (as in Pliny, *Natural History* 9.107, 12.15, 17.132). In that case, he would be correctly representing the problem of the Sicilian sundial in Rome.

The accuracy of a sundial can be checked for its particular locality, because the solstice and equinox lines should correspond to a specific latitude, as we have just seen with the Catania dial, and as I demonstrated earlier for the dial from Alexandria. Given the facts that all ancient sundials were hand-made, and that their geometrical parameters, such as the angle of latitude, might be expressed in simple but practicable ratios, allowance should be made for some inaccuracy in the instruments.[66] According to Gibbs, the lines on the dials from Delos show that most were made to be used there, whereas the dials from Pompeii show far less accuracy and do not suit its latitude well.[67] That the inaccuracy is not necessarily a result of the dials' having been manufactured elsewhere and brought into Pompeii without a care for their proper location is shown by the existence of one made in effect for a latitude near 50°, which was nevertheless inscribed directly on the horizontal face of a garden column in the town.[68] This inaccuracy in Pompeii means that the days of the solstices and equinoxes were wrongly marked. On the other hand, the daily hour lines on Pompeian dials do match reality reasonably well, so it may be that among the inhabitants of Pompeii there was more emphasis on the time of day and on business within each day than on the time of year.

We do not know how the Romans discovered that the Catania dial was inaccurate for them. It is possible that the presence in Rome itself of more and more sundials through the second century provided the observant with evidence that it was indeed unreliable. A passing comment in one of Plautus' plays indicates that the concept of the hour had already made its way to Rome by the early second century BC, and this was probably in the wake of actual Greek sundials. Plautus used the inequality of the seasonal hours to make a joke about the excessive drinking habits of one of his characters (*Pseudolus* 1304): this man is accused of being able to drink an enormous quantity of wine in an hour, to which he responds 'a winter hour, that is', in an effort to lessen the charge by making the time shorter.

One might therefore guess that simple 'macroscopic inspection' revealed the error of the misplaced Sicilian sundial in Rome, but then why did the Romans live with that error for so long? Did the calendrical aspect of the dial mean less to them than its function as a daily clock, as we have surmised for the Pompeians?

Certainly we gain a distinct impression that the hours of the day mattered more to the Romans, as there is a careful parcelling out of them. In Rome, the *accensus* had the job of announcing when it was (the end of) the third, sixth, and ninth hours of the day (Pliny, *Natural History* 7.212; Varro, *On the Latin Language* 6.89). It is interesting to note that the sixth hour, signalling noon, was noted not by the height of the sun, but by a distinctly artificial observation within the built environment in Rome – by the passage of the sun between the Rostra and the Graecostasis in the Forum, when viewed from the Senate House – thus creating a makeshift sundial out of the local architecture. We are not told in the literature how the third and ninth hours were recognised, but since a few sundials do survive with these hours specifically marked out, some such mechanism would seem most likely.[69] The day was thus divided into quarters. These in turn were inherited by the early Christian church, along with the Jewish practice of praying a certain number of times during the day (cf. *Psalm* 119.164, *Acts* 3.1), to form parts of the canonical hours of prayer named (from the Latin) as Terce, Sext and Nones, bounded at one end of the day by Matins (sunrise) and Prime (the first hour), and at the other by Vespers (evening) and Compline (end, or completion, of the day).[70]

But even more precise dividing up of the sacred day can already be seen in early Imperial times in the timetabling of events at certain hours for the Secular Games (*Ludi Saeculares*) in 17 BC:

> . . . on the Nones of June [= 5 June] Latin [plays] in the wooden theatre which is by the Tiber at the second hour; a Greek chorus in the theatre of Pompey at the third hour, Greek stage plays in the theatre which is in the Circus Flaminius at the fourth hour.
>
> (*CIL* VI. 32323. 155–8[71])

In the same period, Horace records how one mark of his friendship with Maecenas is the use of everyday conversational gambits between the two, including the question, 'What hour is it?' (Horace, *Satire* 2.6.44). We noted earlier the practice among the Romans of having a slave check outside for the hour of the day from a sundial, particularly during lengthy court proceedings (Cicero, *Brutus* 200; Juvenal 10.216; Martial 8.67.1).

In addition, the hour lines on dials were sometimes numbered. If this was done in Greek, the letters of the alphabet were used for numbers (see Figure 4.4). A Greek epigram puns neatly: 'Six hours are sufficient for work. But the rest, when set out in letters, say "Live!" to mortals.' (*Anthologia Palatina* 10.43) The first six hours of the day, to noon, were devoted to work, but the next four to leisure. This is a simple play on the letters for the hours 7, 8, 9, and 10, which were Z, H, Θ, and I; read together, they formed the word *zēthi*: 'Live.'

And yet the hour could be destructive too, as Horace reminds us: 'The year warns you, lest you hope for immortality, as does the hour, which seizes the nourishing day.' (Horace, *Ode* 4.7.7–8)

For Seneca it is a mark of the over-pampered leisure-class that they have to be reminded by someone else when to wash, swim or dine (*On the Shortness of Life* 12.6). But underlying this deprecating characterisation of some of his fellow Romans there lies for the first time signs of a carefully scheduled day, for which the satirist Martial (*Epigrams* 4.8) provides sharper definition.[72] Although one can still discern in his timetable the prime demarcation created by the traditional quarters of daytime, the greater precision of his timing must reflect an increased use of sundials. The first two hours were occupied by the *salutatio* between patron and clients; at the third hour the law courts opened; work throughout the city lasted till the end of the fifth hour, followed by a rest at the sixth, and a complete end to work at the seventh; the eighth hour was spent at the gymnasium; dinner came at the ninth; the tenth, the poet hopes, is when the emperor will read this latest book of epigrams(!). These hours may represent the norm for the period or place (barring the very specific use of the last hour), as elsewhere Martial could satirise someone for arriving for dinner too early, before even the fifth hour has been announced, and the fourth is still filled with law court wranglings and the arena is occupied by hunts (*Epigrams* 8.67); and the sixth, seventh and eighth hours are practically synonymous with the baths (*Epigrams* 10.48.1–4).

But variations certainly occurred. Later in the second century, Lucian takes midday ('the *gnōmōn* is shading the middle of the *polos*': *Lexiphanes* 4) as the time for a visit to the baths, where a good work-out at the gymnasium was rounded off by a plunge or swim in the pools. He can, incidentally, alternatively still express this bathing time in a suitably archaistic fashion as a short foot-measure, in a set of mock laws for the celebration of the Saturnalia: 'Bathing is whenever the *stoicheion* is six feet' (*Saturnalia* 17). Contemporaneously, Artemidoros, the dream-interpreter, emphasises the good fortune that lies inherent in the busy-ness of the morning hours, when he informs us:

> The *hōrologion* symbolises deeds and actions and movements and the undertaking of business. For men do everything with an eye to the hours. Therefore, if the *hōrologion* falls or is broken, it would be evil and destructive, especially for those who are ill. It is always better to count the hours before the sixth than those after the sixth.
>
> (Artemidoros 3.66[73])

Pliny the Younger describes his own day in the comfort of his villa in Tuscany (*Letters* 9.36). He would rise more or less around sunrise, but stay in a silent, darkened room to think and compose any writing that was needed. At the fourth or fifth hour, depending on the weather, he would head out for a short amble in the villa, still thinking and composing, and then go for a carriage drive. This would be followed by a sleep, another walk, some declamation in Greek or Latin, yet another walk, and then some exercise and a bath. Supper would be taken, then Pliny would take a final walk with his family.

Retirement, one might have thought, should have brought an even more flexible timetable for the rich, but as Pliny the Younger (*Letters* 3.1) demonstrates in his description of the typical day of the elderly but sprightly Spurinna, it could be even more regimented, with tasks set at specific hours. This 77-year old would stay in bed for the first hour of the day, then at the second hour call for his shoes and go for a three-hour walk. This would be followed (by definition, from the fifth hour) by conversation with friends at home, or listening to a reading, and a rest. Then – surprisingly, during the hottest part of the day – he would go for a seven-mile carriage ride, and another mile's walk, before retiring to bed or to write. Exercise at a ball-game and bathing followed the siesta. Allowance was made for the variability of the seasons for bathing, as the time would be shifted from the ninth hour in winter back to the eighth in summer, presumably so as to shorten the period of pre-bathing exercise over the hottest part of the day in summer. A further spell of listening to a reading would precede dinner, which would last into the night, even on the longer summer nights.

Spurinna's bathtimes in summer and winter are worth analysing for what they illustrate of the differences between ancient and modern perceptions of time. The ninth hour in winter and the eighth hour in summer are quite different distances from sunset: obviously three and four seasonal hours respectively, but in equinoctial hours 2h 15m short of sunset in winter and 5h 3m shy in summer. The sun is also at distinctly different altitudes: 17°19′ in winter, but 53°27′ in summer. On the other hand, these bathtimes lie at fairly similar temporal distances from midday in terms of equinoctial hours: 2h 28m in winter, and just 2h 40m in summer. What this suggests is that it was not the sun's heat which was prioritised for the bath – this would have been a reasonable notion if Spurinna's private bath took advantage of the sun's natural heat in addition to that of a furnace. Nor was the sun's altitude a signal for exercise and bathing, since there is a considerable difference between the winter and summer observations in this regard. Nor was the end of the day the definer of bathtime, since the activity lay at quite different distances from it in both seasonal and equinoctial terms. Rather, it was midday, with its accompanying outdoor activity and the subsequent siesta, which signalled when Spurinna should start looking towards his exercise and bath. Although in Spurinna's mind the bath-related activities took place a whole hour earlier or later at the seasonal extremes, to our minds they would occur at much the same time, more or less two-and-a-half (60-minute) hours after midday. In our terms we would exercise and rest for much the same period of time whether it was winter or summer, whereas to Spurinna's way of thinking, he would expend an hour's less energy in the heat of summer than he would in winter. Our standardisation of time suggests that there is an underlying, biological sensitivity to light and heat – a circadian rhythm of sorts – which might have been controlling Spurinna's daily physical activity (or lack of it). Even his regular three-hour walk will necessarily have varied in length in our terms through the seasons. It is possible, though unprovable, that Spurinna measured such an activity through some type of portable sundial, but further research is needed on people's ability to measure the passage of seasonal hours without artificial aids.

All this marking of time through the day signifies the increasing tendency among the élite under the Empire to consult the sundial, at least for major demarcations of the passage of time. But then, as now, keeping to a regular timetable was also of concern to those involved in the delivery of messages or other mail. There were in the classical world special classes of runners, whose names usually were signifiers of the linear measure that they ran. The *stadiodromos* (Pindar, *Olympian* 13; Plato, *Laws* 833a7–10), for instance, obviously ran the *stadion* itself, or about 1,828 metres, while the *dolichodromos* (Plato, *Protagoras* 335e4) was named after the distance run over the *dolichos*, a length of 20 stades, or about 36.57 km.[74] The original Marathon runner, Pheidippides, on the other hand, who extraordinarily ran from Athens to Sparta and back again (Herodotos 6.105, cf. Cornelius Nepos 4.3), was a *hēmerodromos*, a 'day-runner', a name noteworthy in this context for its unique, explicit expression of time rather than distance run. While the ultimate *hēmerodromos* was the sun (cf. *Suda*), Livy usefully defines the term for his Roman readers as one used by the Greeks to refer to 'those who measure out a huge distance by running in one day' (Livy 31.24). The notion of running from Athens to Sparta in a day is no fantasy: the modern 'Spartathlon', trialled in 1982 and officially run annually since 1983, seeks to replicate the original 'Marathon' by covering the 250 km from Athens to Sparta within 36 hours; the record, set in 1984 by Yiannis Kouros, is 20 hours 25 minutes.[75]

The name *hēmerodromos* is applied to couriers generally into the Roman period: Strabo records how in South Italy at the end of the third century BC:

> Picentia used to be the mother-city of the Picentes; but now they live in villages, after being been expelled by the Romans because of their alliance with Hannibal. And instead of military service, they were appointed then to be public day-runners (*hēmerodromein*) and letter-carriers, just like the Leucani and Bruttii for the same reasons.
>
> (Strabo 5.4.13)

Aulus Gellius (10.3.19) relates much the same, but makes it plain that this alternate service imposed on the pro-Hannibal peoples of the south was a humiliation.[76] The system of *hēmerodromoi* as messengers is thought to have continued in Macedonia and Greece until the Roman period.[77] It is not apparent that these particular 'day-runners' worked in relays – clearly, Pheidippides did not on his famous run, and probably between most Greek cities there was no great need to do so – but in other areas of the Hellenistic world and in the Roman Empire, relay-runners or riders certainly worked for what we may rightly call a postal service, albeit one limited to top government officials. To the managers of these services the ability to account for not just the day but even the hours within the day was sometimes necessary.

The sending and receipt of letters in a world without a centralised or state, communal, postal service was a haphazard affair, with issues surrounding the adoption of a suitable messenger, means of travel, security en route, the lack of formal street addresses and so on. All the same, such a system seems to have satisfied most customers most of the time, to judge from the surviving letters of both the well-known and the anonymous. Naturally, the wealthy had the advantage that they could assign slaves or freedmen to the task of carrying letters from one place to another, but anyone could ask a friend or even a stranger who was travelling in the right direction to deliver a letter. Even more remarkable is the fact that such letters actually found their way to their intended recipients. How long it took to receive private letters seems to have been less of a concern to some than whether a letter was sent at all. Inveterate letter-writers like Cicero would send a letter every day if they felt like it, despatching them by whatever means available. In his case, timing was not a concern unless news of significant political or family events was sought.[78]

Nonetheless, in this generally free-for-all world, there were some formal services provided by and for the state in different places and times. The Persian Empire established its own horse-borne postal service, which was adopted by Alexander and his successors following his conquests. Ptolemaic Egypt inherited its service from the pre-existing Persian operation, and it is from Egypt that we have records which provide evidence of the timing of the service according to the hours, as the following extract from a papyrus dated to 259–253 BC illustrates:

> At the first hour Theochrēstos delivered from upper [Egypt] to Dinias three rolls, [of these] for king Ptolemaios two rolls; for Apollonios, the controller, one roll, and Dinias delivered [them] to Hippolysos.
>
> At the sixth hour Phoenix, son of Herakleitos the elder, Macedonian (holder of a hundred aroura), of the Herakleopolite (nome), . . . delivered one roll to Phanias, and Aminon delivered (it) to Timokrates.
>
> At the eleventh hour Nikodemos delivered from lower (Egypt) to Alexandros . . . rolls, from king Ptolemaios for Antiochos in the Herakleopolite (nome) one roll; for Demetrios, (who is) in charge of the supply of the elephants, in the Thebaid, one roll; for Hippoteles, who is with Antiochos against Andronikos, in Apollonopolis Magna one roll; from king Ptolemaios for Theogenes, the money-carrier, one roll; for Herakleodoros in the Thebaid [one roll]; for Zoilos, the banker, of the Hermopolite (nome) one roll; for Dionysios, the manager, in the Arsinoite (nome) one roll.
>
> (*P. Hibeh* 1.110, 65–87[79])

This is part of a daily logbook kept by an official probably called a *hōrographos*, an 'hour-writer', to judge from similar documents from elsewhere in Egypt.[80] It comes from a single postal station in a relay of such stations, which operated by land along the Nile to and from Alexandria. Analysis of the surviving entries has suggested that the journey between relay stations on the route was undertaken by horse, took six hours, and continued overnight; that there were four deliveries through this particular station in the daytime – from south to north at the first and twelfth hours, and from north to south at the sixth and twelfth hours; that this logbook records only items which needed to be sent by express post between the central government and local officials; that other pressing letters between local officials probably made use of the same service, but the passage of their correspondence was recorded perhaps in another logbook; and that less important mail would have been sent by other means. Llewelyn, however, has argued that much of this picture of postal punctuality is largely supposition, based on descriptions of the earlier Persian relay system, known as *angarēion* (Herodotos 8.98; Xenophon, *Cyropaedia* 8.6.17). In reality, he believes, the Ptolemaic service may have been less rigidly timetabled than has been supposed.[81] Nevertheless, Llewelyn's belief that the hours enumerated in the logbooks are simply rounded up to whole numbers – with the first hour standing for any time around sunrise, the sixth hour for around midday, and the twelfth hour for about sunset – has rightly been disputed by Remijsen. She points out that since the other two hours enumerated in the surviving logbooks, the *hōra heōthinē* (the 'hour of dawn', hence the twelfth hour of the night) and the eleventh hour are precise, then the first, sixth and twelfth probably are so too.[82]

In the Roman world, Augustus introduced in two phases a system for express post for official correspondence, which lasted through the Empire.[83] Initially he used relays of couriers, but then, prioritising accuracy of reporting over speed of delivery, he changed the system to one of relays of carriages, so that the same single messenger made the whole journey and was available to answer questions at his destination (Suetonius, *Augustus* 49.3).[84] By chance, we have from Egypt the Roman equivalent of the Ptolemaic logbooks, preserved on *ostraka*, throw-away fragments of pottery, from the military camp at Krokodilo in the eastern desert between the Nile and the Red Sea in southern Egypt. Cavalry from the camp served as occasional couriers and escorts, who on arrival at Krokodilo reported to the camp's *curator* what he was delivering. The *curator* recorded this on a handy piece of pottery (which could be disposed of afterwards when the record was no longer needed), and passed the post on to another messenger for the next stage. These couriers returned to their home camp immediately, just two or three hours' journey away, without necessarily awaiting any return mail. The service ran both day and night, and again the hours enumerated are precise – down, in fact, to the half-hour: 'Letters came for me at the eleventh and a half hour of the day and left at the twelfth hour.'[85]

Although the Ptolemaic and Imperial Roman services appear to have been *ad hoc*, it is apparent that there was the capacity within both postal systems for carefully timetabled expectations of speed or punctuality. Our awareness of this potential is a result of the purely accidental recovery of records on Egyptian papyri and pottery. Knowledge of such 'hour passes' as are recorded in Egypt was, it seems, lost under the later Roman Empire, to judge from the fact that medieval Europe rediscovered them only after hearing of such things in the Asian postal services encountered by travellers like Marco Polo.[86]

While great speed could occasionally be achieved, especially on the Roman postal routes, as we have seen speed was not of the essence with their service, and its effects appear not to have been recorded.[87] It is a pity in the case of the Egyptian services, the ancient postal route which currently provides us with the most information about the timing of travel between the postal stations, that the lengthy routes ran more or less north–south along the Nile, while the east–west routes through Krokodilo were very short. Had there been a similar service which ran a considerable distance at speed from east to west and back, then we might have encountered some expression in the documentation of the effects of fast travel across several towns' 'time zones'. There was certainly a recognition in antiquity that the sun rose and set at different times over what we call different longitudes, so that dawn would occur earlier in Persia than it would in Spain – this is one of the arguments adduced to prove the sphericity of the earth by Kleomedes (I.5.30–57). The recording of the same lunar or solar eclipse in different locations confirmed the time difference, a notable instance being a lunar eclipse observed at Arbela in northern Iraq at one extreme and in Sicily or at Carthage at the other on 20 September 331 BC, a few days before Alexander's battle against Darius III at Gaugamela.[88] Speedy travel across the 'time zones' would have provided a further illustration of this phenomenon but on a more human and less spectacular level than the celestial phenomena provide. It will take the development of 'high speed' travel and communication technology in the nineteenth century to spur nations into the establishment of national standard times and international time zones, in order to facilitate the timetabling and coordination of long-distance express travel and communications.[89]

But another modern convention, of measuring great distances by duration, such as light-years in astronomy, clearly has its ancient counterparts. The conventional means of measuring distance was by the linear space traversed, as we have seen embedded in the names of Greek runners, such as the *stadiodromos* and the *dolichodromos*. Linear distance is also the explicit measurement on Roman route maps, such as the 'Antonine Itinerary' of the third century AD and the twelfth-century 'Peutinger Table', which may ultimately derive from a second-century AD original, as well as of the Vicarello cups, a set of four silver cylinders from northern Italy, dating to the first two centuries AD and engraved with lists of all the post-stations on the itinerary from Cadiz in Spain to Rome, and the distances between the posts.[90] But we have also seen that time could be used to denote distance, in the name of the long-distance runner, the *hēmerodromos*, a 'day-runner'.

This is not the only case of such usage, as we find it also in travel by sea. Herodotos gives an early instance. After giving the dimensions in Greek *stadioi* of the Pontus, the Bosporus, the Propontis and the Hellespont, he explains the method of calculation:

> These measurements have been made in this way: a ship will generally accomplish 70,000 *orguiae* in a long day's voyage, and 60,000 by night. This being granted, seeing that from the Pontus' mouth to the Phasis (which is the greatest length of the sea) it is a voyage of nine days and eight nights, the length of it will be 1,110,000 *orguiai*, which make 11,000 *stadioi*.
>
> (Herodotos 4.86.1–2)

The conversion from duration to distance means that one day's voyage comprises almost 700 *stadioi*, and a night's voyage practically 600 *stadioi*. A full 24-hour period, the *nychthēmeron*, is later equated with 1,000 *stadioi* (pseudo-Skylax). The practice has a long history, arguably even forming the basis of calculations of the cost of sea transport in the emperor Diocletian's Edict on Prices of AD 301.[91] Time truly does come to mean money.

7

EPILOGUE

In Chapter 1 we saw briefly how the natural landscape could be used to assist in marking time through the year, whether that was the natural, seasonal year or an artificial year, like a religious, festival year. In particular, I pointed out how in ancient Athens the summer solstice sunrise could be marked against the conical peak of Mount Lykabettos when viewed from the political centre of the city, the Pnyx. That period of the year mattered culturally, because the festival calendar began its New Year with the first new moon following the summer solstice. And that calendar in its turn governed to some extent when political meetings could or could not be held. In this chapter I want to examine briefly the use of the built environment for telling the time in antiquity.

When we dealt with sundials in the previous chapters, I noted the three principal types used in antiquity – the spherical, the conical and the plane. There are, indeed, various subtypes of these, and one in particular deserves mention as I think it leads to a better understanding of the use of time in one of the major pieces of surviving ancient architecture, the Pantheon in Rome. This gives us one final, very large-scale example of the cultural uses of timekeeping.

If a sundial is a spherical one, like the Roman example from Aphrodisias in Turkey in Figure 4.4, the carved interior recreates in mirror-image the apparent hemispherical dome of the sky. For economic reasons the hemisphere is often cut down to just a quarter-sphere, and captures only that part of the sky occupied by the sun in the course of the year. The dial typically lets its *gnōmōn*'s shadow fall on to a series of curved lines incised on the interior of the open quarter-sphere. These lines represent the daily passage of the sun, usually on four particular occasions but utilising only three lines: the summer and winter solstices on separate lines at the extremes of the dial, and the two equinoxes, using the same single line between the solstices. Figures 7.1 to 7.3 represent a section through the meridian (noontime line) of a hypothetical spherical sundial, with the line of the sunlight and the resultant shadows from the *gnōmōn* for these major seasonal points.

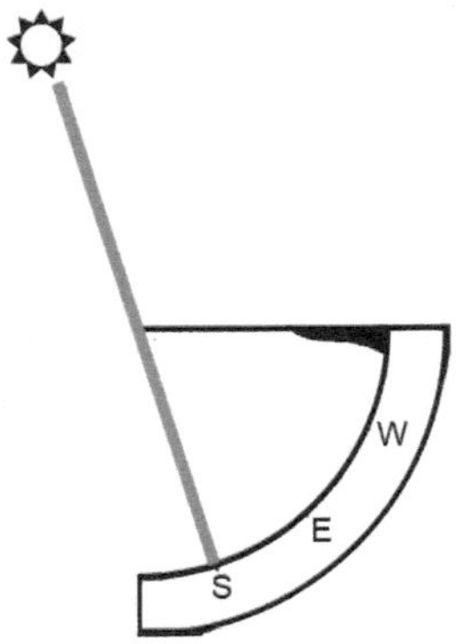

Figure 7.1 Profile through the meridian of a spherical sundial, showing the *gnōmōn*'s shadow at the summer solstice.

Source: Drawing by R. Hannah.

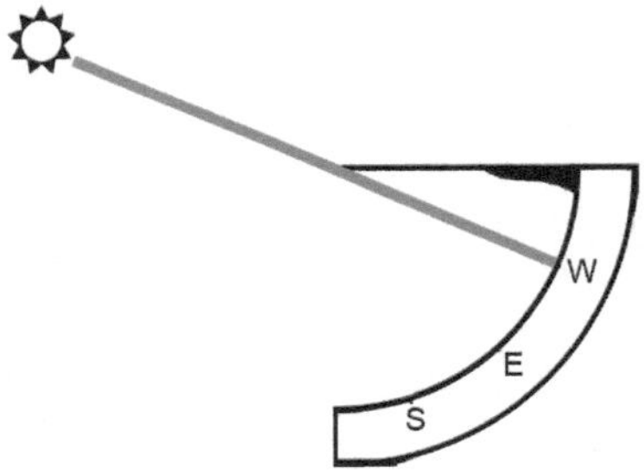

Figure 7.2 Profile through the meridian of a spherical sundial, showing the *gnōmōn*'s shadow at the winter solstice.

Source: Drawing by R. Hannah.

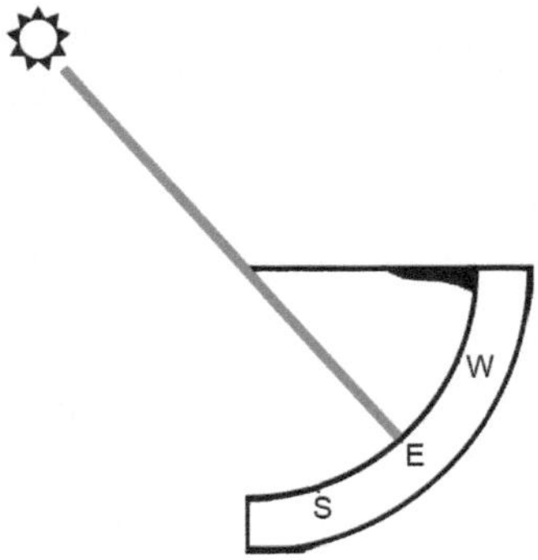

Figure 7.3 Profile through the meridian of a spherical sundial, showing the *gnōmōn*'s shadow at the spring and autumn equinoxes.

Source: Drawing by R. Hannah.

The North Celestial Pole is perpendicular to the equator, and therefore to the equinoctial line. Since the angle of the North Celestial Pole above the horizon is the same as the angle of local geographical latitude, from the equinoctial line of the sundial we can calculate the dial's intended latitude, as in Figure 7.4. Here the latitude turns out to be 43°N.

There were also, however, 'roofed' spherical sundials which captured the sunlight itself within a shadowy interior.[1] They consisted of a stone block carved out into a hollow hemisphere, with a hole let into its upper surface. Through this hole the sunlight filtered on to the engraved surface inside. Once again, we can use a hypothetical example to illustrate the principle, taking a cross-section through the meridian, as in Figures 7.5 to 7.7.

The latitude is again given by the equinoctial line, in this case 45°N (Figure 7.8).

If we shift now from these practical sundials to a monument much larger in scale, but, I believe, closely related in form to the roofed sundial, we will see how it too might be used as a timing device. This is the Pantheon in Rome (Figure 7.9).

The first version of this building was constructed by Augustus' general, Agrippa, in 27 BC, but destroyed by fire in AD 80 and then restored by the emperor Domitian. It was burned again by lightning in the time of Trajan, and rebuilt from the foundations upwards in its present form, perhaps soon after AD 110 under Trajan, and finally completed by AD 128 by his successor Hadrian.[2] It is a huge domed structure, over 43 m in height, which lets light into the interior through a large, 9 metre-wide *oculus* in the centre of its roof. The temple's form is essentially that of a sphere with its lower half enclosed within a cylinder. It is arguably a close relative in geometrical terms to a roofed sundial (Figure 7.10).

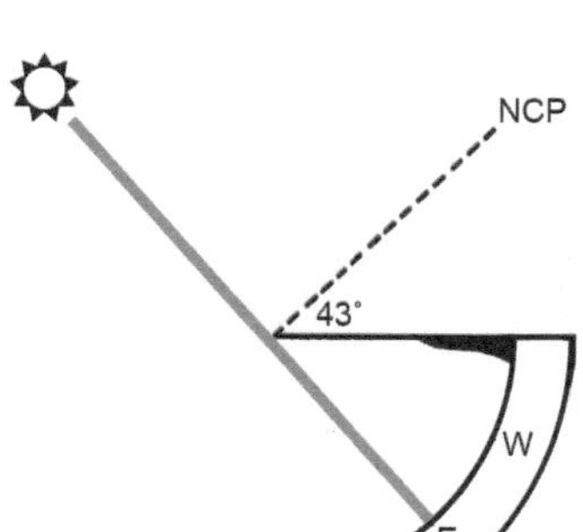

Figure 7.4 Profile through the meridian of a spherical sundial. Determination of the latitude from the equinoctial line. NCP = North Celestial Pole, latitude = 43°N.

Source: Drawing by R. Hannah.

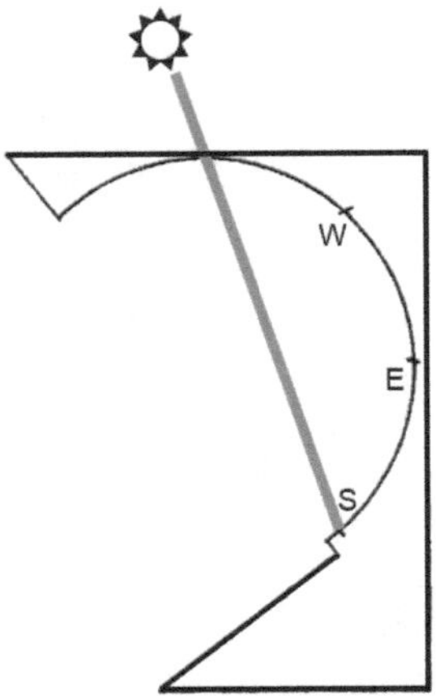

Figure 7.5 Profile through the meridian of a roofed spherical sundial, showing the beam of sunlight at the summer solstice.

Source: Drawing by R. Hannah, after Gibbs 1976: 23 Figure 9.

The Pantheon faces almost directly north, so that the sunlight, coming in from the south, falls over the northern entranceway at noon. The temple draws attention to two solar phenomena, both dynamic through the year: the midday sun, and the sun's daily path. I want to concentrate on the first.

The sun spends six months of the year, in spring and summer, falling below the level of the dome.[3] Figure 7.11 illustrates the situation at the time of the summer solstice. This is when the sun's altitude is at its highest at noon for the year; for the latitude of Rome this altitude is 72°.

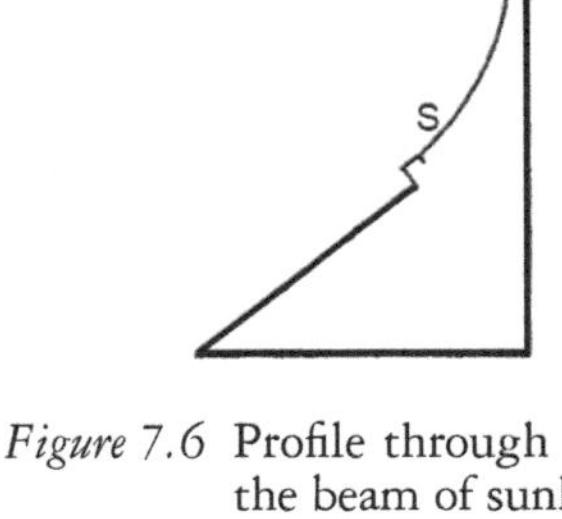

Figure 7.6 Profile through the meridian of a roofed spherical sundial, showing the beam of sunlight at the winter solstice.

Source: Drawing by R. Hannah, after Gibbs 1976: 23 Figure 9.

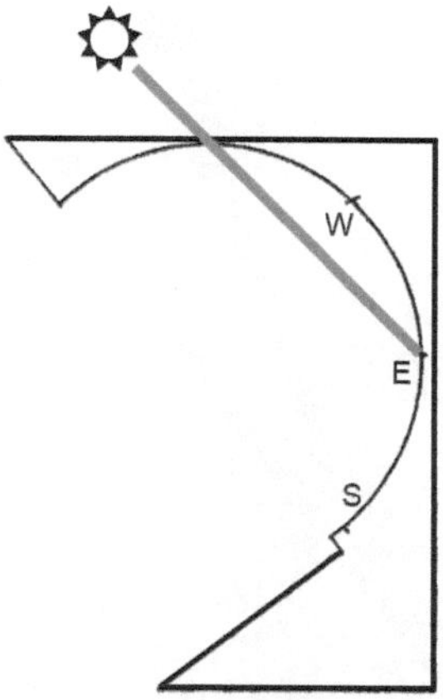

Figure 7.7 Profile through the meridian of a roofed spherical sundial, showing the beam of sunlight at the spring and autumn equinoxes.

Source: Drawing by R. Hannah, after Gibbs 1976: 23 Figure 9.

The sun then spends six months, in autumn and winter, shining on the interior of the dome. Figure 7.12 illustrates the period of noon at the time of the winter solstice. This is when the sun's altitude is at its lowest at noon for the year: in this case at 24°.

The shift from one half-year to the other is marked by the passage of the sun at the equinoxes in March and September. At this point the midday sun shines partially just below the dome, to pass through the grill over the

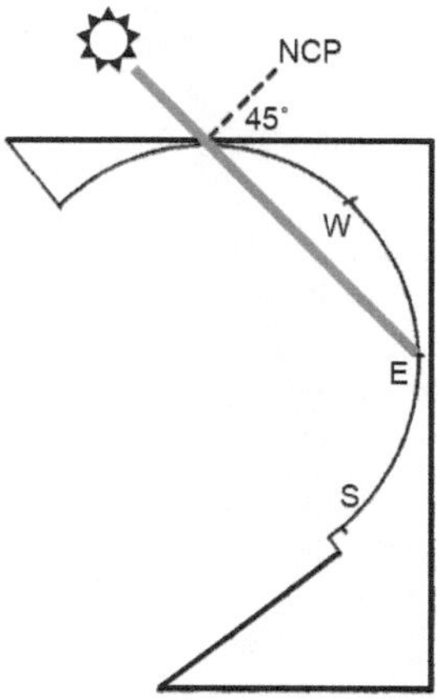

Figure 7.8 Profile through the meridian of a roofed spherical sundial. Determination of the latitude of a spherical sundial from the equinoctial line, in this case 45°N.

Source: Drawing by R. Hannah, after Gibbs 1976: 23 Figure 9.

Figure 7.9 Rome, Pantheon.

Source: Photograph reproduced by kind permission of N. Hannah.

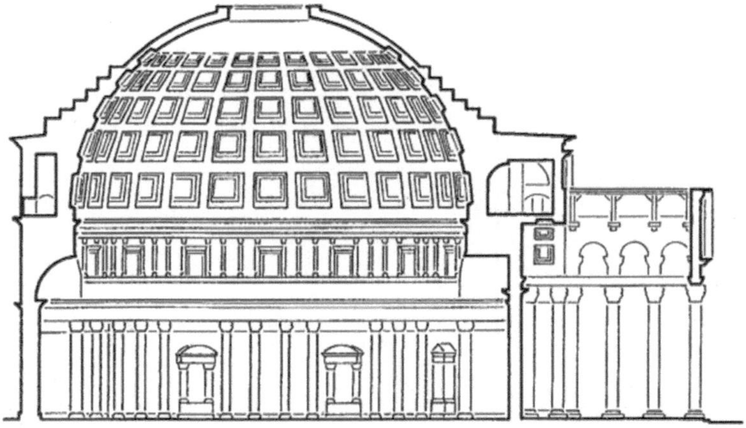

Figure 7.10 Rome, section through the Pantheon.

Source: Drawing from Ward-Perkins 1979: 87 fig. 127; reproduced by kind permission of Electa, Milan.

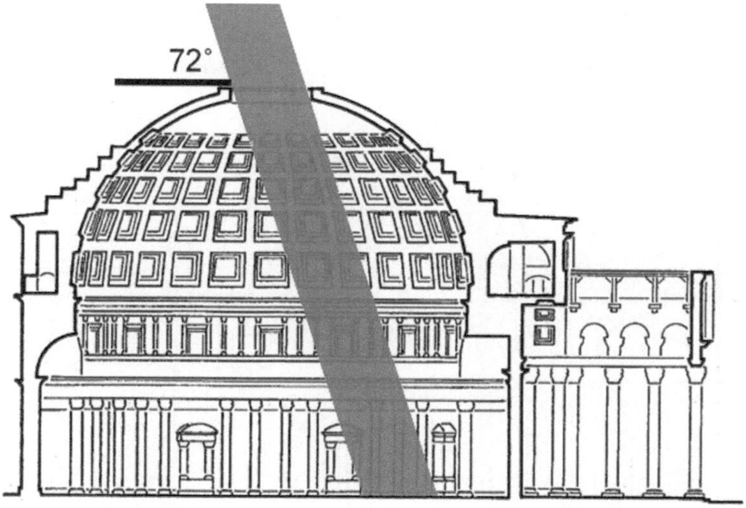

Figure 7.11 Rome, section through the Pantheon. The shaded area shows the sunlight through the *oculus* at noon on the summer solstice, when the sun's altitude is 72°.

Source: After a drawing from Ward-Perkins 1979: 87 fig. 127; reproduced by kind permission of Electa, Milan.

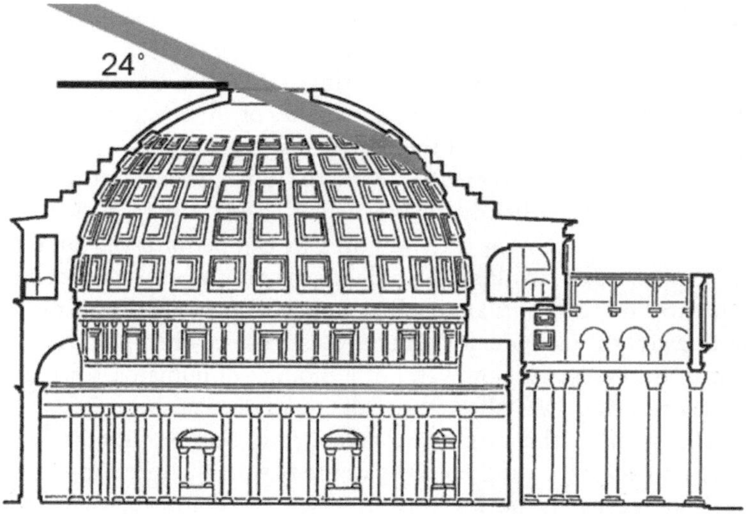

Figure 7.12 Rome, section through the Pantheon. The shaded area shows the sunlight through the *oculus* at noon on the winter solstice, when the sun's altitude is 24°.

Source: After a drawing from Ward-Perkins 1979: 87 fig. 127; reproduced by kind permission of Electa, Milan.

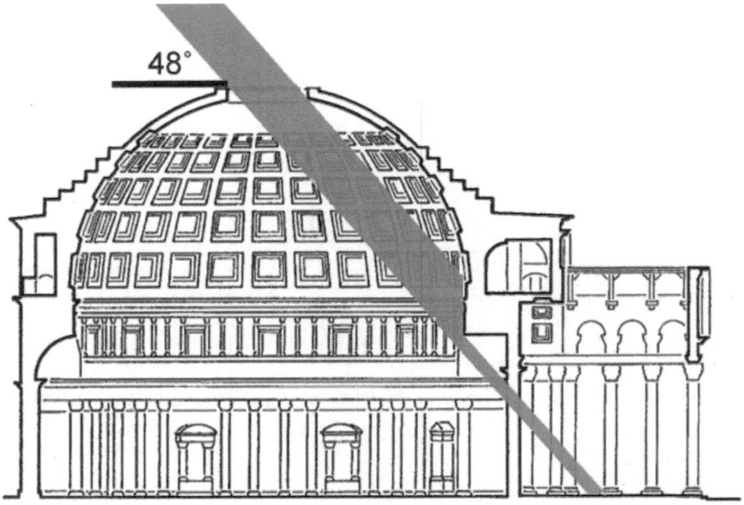

Figure 7.13 Rome, section through the Pantheon. The shaded area shows the sunlight through the *oculus* at noon on the equinoxes, when the sun's altitude is 48°.

Source: After a drawing from Ward-Perkins 1979: 87 fig. 127; reproduced by kind permission of Electa, Milan.

entrance doorway and fall on the floor of the porch outside. Figure 7.13 represents this period, when the sun's altitude is about 48°.

The centre of this equinoctial, noontime circle lies on the architectural moulding, which marks the base of the dome, or, in effect, the diameter of the imaginary interior sphere of the temple (Figure 7.14).

The *oculus* in the dome, therefore, serves the same purpose as the hole in the roof of the spherical sundial. Its lip acts as the *gnōmōn*, and from that and from the angle of the sunbeams at the equinoxes we can perform the same calculations to determine the working latitude for the Pantheon as a sundial (Figure 7.15). It turns out that this angle of latitude is 42°, which is, unsurprisingly and necessarily, the latitude of Rome, because that is where the building is.[4]

What is not necessary, however, but a product of human artifice, is where the equinoctial light was allowed to fall, since the Pantheon is not entirely spherical in form. In this case it was a deliberate choice of the designer that this light falls on the ceiling precisely at the base of the interior of the hemispherical dome, that is, on its equator, where the dome appears to end and the cylinder begins. This 'equator', however, is entirely illusory in structural terms, for while the dome appears to spring from this point on the interior, in fact it does not. Instead, it springs from around the level of the second row of coffers above the 'equator', as can be seen from the exterior of the dome. So the base of the interior of the dome has been emphasised for some reason.

Figure 7.14 Rome, Pantheon. Sunlight at noon, around the autumn equinox (23 September 2005).

Source: Photograph R. Hannah.

The architect's orchestration of sunlight and structure inside the Pantheon demands explanation. It seems to me that the explanation lies in thinking of the Pantheon as a form of roofed sundial. What sort of time it kept, though, is not obvious. It seems likely that there is a correlation between the architecture and the times when the noontime sun entered the signs of the zodiac through the year, so that the coffers of the ceiling and the marble paving slabs of the floor provide markers for this passage of the sun.[5] Nevertheless nothing in the interior decoration of the building now draws attention to this annual noontime line. If any decorative scheme on the ceiling did so in the past, it is now unfortunately lost. In addition, my association between the Pantheon and a roofed spherical sundial loses coherence outside the midday period, since the coffers continue horizontally from their noontime point above the doorway, not in parabolic curves as they would if they were

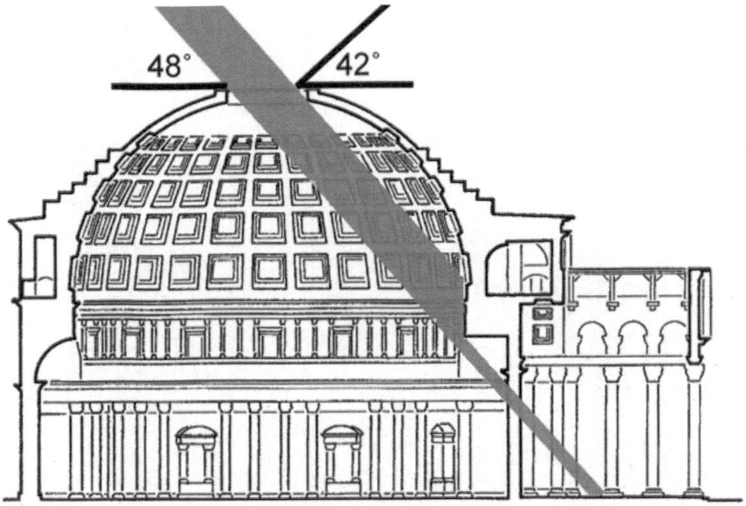

Figure 7.15 Rome, section through the Pantheon. The shaded area shows the sunlight through the *oculus* at noon on the equinoxes, when the sun's altitude is 48°. Determination, from the equinoctial line, of the North Celestial Pole at altitude 42°.

Source: After a drawing from Ward-Perkins 1979: 87 fig. 127; reproduced by kind permission of Electa, Milan.

really part of a gigantic working sundial. We might imagine the doorway coffers forming an imaginary meridian line leading up towards the *oculus*, but that is all it can be, imaginary, because nothing seems to have marked this line of coffers out as anything special.

Yet there is no denying that the equinoctial sun's noontime light does draw attention to the base of the dome. It effectively defines the structure's articulation between sphere and cylinder. Why it does so might be discovered in another Imperial building from an earlier generation. Nero's palace in Rome, the Golden House, built in AD 64–8, contains a large, domed room, known as the Octagonal Room because of its eight-sided ground plan. It is aligned, like the Pantheon, along a north–south axis, so that the equinoctial midday sun falls directly on to the north door of the room. The lower rim of this sun's circle strikes the juncture of the floor and the northern doorway's threshold, which lies on the perimeter of the room. So the sun in effect marks and measures out the dimensions of the room.[6] Once again, as with the Pantheon, we are drawn to the equinox as the marker for the dimensions of the structure, but this time not at the base of the ceiling dome but at the floor level. Why should the equinoctial moment be chosen to provide a basic module of the design?[7]

Nero's association with the sun is well-attested. In the Golden House complex there was a colossal statue of the Sun god, which stood not far from the Octagonal Room (it gave its name eventually to the Colosseum, which was later built nearby). In Nero's later portraits he wears the radiate crown usually associated with the Sun god.[8] And Suetonius reports of the Golden House itself that it had astronomical associations:

> He built a house from the Palatine all the way to the Esquiline, which he called the Passageway House at first, but then, when it was destroyed by fire soon afterwards and rebuilt, the Golden House . . . The main dining hall was circular; it turned round constantly day and night, like the heavens.
>
> (Suetonius, *Nero* 31)

The contemporary poet Lucan, forced to commit suicide by Nero in the years when the Golden House was being built, addresses the emperor at the start of his poem, *The Civil War* (1. 45–62). He advises Nero to find rest, in apotheosis, neither in the northern half of the heavens nor in the southern, but midway. The poet therefore reserves the celestial equator as the one stable part of the sky where the emperor Nero should reside, there to guarantee peace on earth.

Both the Golden House's Octagonal Room and the Pantheon make play with the equatorial, or more specifically equinoctial, part of the sky. In both it helps to define the interior dimensions and structural form. In antiquity, the dome was compared to the celestial dome.[9] The name 'Pantheon' means 'all the gods', and on the basis of the Pantheon's domed form the historian Cassius Dio (53.27) interpreted the temple as a representation of the sky, where all the gods dwell.[10] Through the play between sunlight and structure, the architects of both the Octagonal Room and the Pantheon sought to raise their emperors above the ordinary and into the immortal company of the gods. In these buildings, the barrier between time and eternity is dissolved, at least for the emperors. One of the means by which this assimilation between time and the non-time of eternity was achieved was by basing the structure of these buildings on the roofed spherical sundial. The Octagonal Room, and hence the Pantheon, thus told the time of peace, stability and, paradoxically, eternity.

Such a play between physical architecture and metaphysical belief was to have a long history. Byzantine architects utilised the sunlight to emphasise crucial parts of the Christian liturgy in their domed churches.[11] Renaissance and Early Modern astronomers designed meridian lines within churches, so as to resolve the disjuncture that had developed between the pivotal religious festival of Easter and the Julian calendar.[12] And perhaps most strikingly, an Australian architect, Philip Hudson, influenced by one of these church meridians, designed the Shrine of Remembrance in Melbourne in 1934 so that sunlight would strike the Stone of Remembrance at the eleventh hour of the eleventh day of the eleventh month, thus commemorating the war dead from the First World War.[13] The sacrifice of the fallen – the light strikes the phrase 'Greater Love Hath No Man' – was to be remembered in perpetuity through the capture of the sun inside the building at a very particular moment.[14] This idea of commemoration in perpetuity is at the core of Anzac ceremonies in Australia and New Zealand, when the well-known verse of Laurence Binyon, from the first months of the Great War, is read out:

> They shall grow not old, as we that are left grow old;
> Age shall not weary them, nor the years condemn.
> At the going down of the sun and in the morning
> We will remember them.
>
> (Laurence Binyon, *For the Fallen* (1914))

Commemoration in perpetuity, or at least so long as the sun rises and sets: what was once the preserve of autocratic Roman emperors has been democratised, albeit at a great cost.

The idea is embodied in a different form in Sydney, but it is a form that brings us back to the use of natural landscape to signal time. At one end of Anzac Bridge near Sydney Harbour stand two colossal statues of Anzac soldiers, one an Australian, the other a New Zealander. The New Zealander faces east, while his Australian comrade-in-arms faces west. Thus the Australian soldier faces the western horizon and hence the setting sun, while his companion faces the eastern horizon and hence the morning sun. Together they represent the same notion of perpetual commemoration that Binyon sought to evoke in poetic form: as long as the sun rises and sets, so long will we remember. That we still think of the sun rising and setting is testimony indeed to the power of the ancient mode of thinking.[15]

NOTES

1 TIME IN ANTIQUITY: AN INTRODUCTION

1 Cf. U. Eco in Lippincott 1999: 10. A useful and brief introduction to the development of the 'Big Bang' theory is given by Hoskin and Gingerich 1997: 355–58.
2 The top-ten most common nouns made it into the popular media, e.g.: http://news.bbc.co.uk/2/hi/uk_news/5104778.stm (accessed 20 June 2008). This list is: 1. Time, 2. Person, 3. Year, 4. Way, 5. Day, 6. Thing, 7. Man, 8. World, 9. Life, 10. Hand.
3 Harvey 2004: 122–23 no. 89. I am grateful to Dr Jon Hall for this and the following references.
4 Harvey 2004: 180–81 no. 159.
5 Bureau International des Poids et Mesures: http://www.bipm.org/en/si/si_brochure/chapter2/2-1/second.html (accessed 20 June 2008).
6 Cf. Hannah 2006a.
7 For a brief survey of the development of a sociology of ancient time, see Hannah 2008.

2 COSMIC TIME

1 Cf. Bowen and Goldstein 1988: 72–77; Hannah 2005: 52–55; Turner 1989: 310–11.
2 Parsons 1936: 1–82, especially 30–31, 61–63.
3 Stephen's sketch (from Parsons 1936: Map 4) captures well what is visible with the naked eye: cf. McCluskey's photograph. For Hopi astronomy in general, see McCluskey 1977.
4 Stephen's azimuths for the solar observations are several degrees out from their actual distance from true north because of the shift of magnetic north from true north which applies at any given time. Magnetic declination – the angle between magnetic north, as read by a compass, and true north – for his position in 1892 was about 14° (http:www.ngdc.noaa.gov/seg/geomag/jsp/USHistoric.jsp; accessed 8 September 2007). Therefore Stephen's reading of 226.3° on 21 December should read 240.3° ± 0.5°.
5 Henriksson 2007. See also: http://www.mikrob.com/seac2001/tombs.html (accessed 16 June 2008).
6 See now Boutsikas 2007, and Salt and Boutsikas 2005 for temples in mainland Greece.
7 See, for instance, the New Zealand government's Ministry for Culture and Heritage website's entry at: http://www.teara.govt.nz/EarthSeaAndSky/

Astronomy/MatarikiMaoriNewYear/en (accessed 10 June 2008). The standard reference work on Māori astronomy is still Best 1922 (available online from the New Zealand Electronic Text Centre at: http://www.nzetc.org/tm/scholarly/tei-BesAstro.html; accessed 10 June 2008). In Best 1922: 38, the heliacal rising of the Pleiades (i.e. at dawn) marked the New Year for tribes on the 'eastern coast of the North Island', whereas other tribes, 'notably the Ngāpuhi district and the Chatham Islands' took the 'cosmic rising of Rigel' as the marker. The Ngāpuhi occupy part of the far north of the North Island, while the Chatham Islands lie some 800 km east of the South Island; the context makes it clear that the dawn rising of Rigel is meant. Dr Jim Williams, of Kāi Tahu descent in the South Island, tells me that Puaka (i.e. Rigel in Orion) is the harbinger of the New Year for southern Māori, rather than the much fainter Matariki (Pleiades); and proposes that Matariki was introduced, along with the kūmara (sweet potato), in a second wave of Māori immigration from East Polynesia to New Zealand; see Williams (forthcoming).

8 The naturally low visibility of the Pleiades, and the low altitude to which they rise in New Zealand's latitudes – they skirt above the horizon, rising to a maximum alitude of only 19°–31° above it – suggest that this celestial marker for the start of the Māori year was initially chosen elsewhere in the Pacific, where these stars are more obvious. For example, in the eastern Pacific in the Marquesas in French Polynesia, the general region where the Māori are thought to have originated, the Pleiades rise much more steeply, and therefore more obviously, to an altitude of 57°. Readers in, say, London will find the situation similar, with these stars rising high to an altitude of 63°. In these circumstances, the Pleiades are considerably easier to identify. The choice of Rigel (Puaka) instead of the Pleiades (Matariki) by some Māori tribes makes a good deal more sense as a marker for New Zealand, since this star is considerably brighter and more obvious; the curiosity is that these tribes reside across the country, and not just in the south where the Pleiades are less visible.

9 The term 'heliacal' derives from the Greek *hēlios* (sun); 'acronychal' signifies 'the edge of night', from the Greek *akros* (extremity), and *nyx* (night), and technically may refer to either the first or last evening sighting; it is sometimes misspelled as 'acronical' (e.g. Bickerman 1980: 112–14).

10 'Cosmical' derives from the Greek *kosmos* (world, universe), so the term is not self-defining in this context.

11 Bobrova and Militarev 1993; Koch-Westenholz 1995: 163–64; for earlier discussions, see van der Waerden 1952–53.

12 It is worth remembering that to the ancient mind astronomy and astrology were two closely related and legitimate avenues to knowledge: as Ptolemy makes clear, predictive power comes from astronomy via the motions of the celestial bodies, and from astrology via the influence of the relative configurations ('aspects') of these same bodies (Ptolemy, *Tetrabiblos* 1.1); see Barton 1994; Beck 2007: 3–8; Taub 1997.

13 Cf. West 1991: 202–5, justifying the use of both this 'Tropical Zodiac' and the constantly updating 'Sidereal Zodiac', which produces the well-known 'Ages' (such as the 'Age of Aquarius'). On the 'Ages', see Parker and Parker 1991: 24–25, in the context of a practising astrologers' handbook.

14 See Beck 2007: 23–25 on the phrase 'is in'.

15 See, in general, McCluskey 2000.

16 Cf. Davidson 1985: 69.

17 See Evans 2002 for a brief but useful description of the events during the Mysteries.

18 West 1978: 347–48.
19 Taverner 1918.
20 For the date of Geminos, see now Bowen 2006: 199–200 n. 4.
21 For a fuller discussion, see Hannah 2005: 29–35.
22 See further Hannah 2005: 34–35.
23 Michels 1967: 131–32.
24 See http://www.hexhamcourant.co.uk/1.21033. I am grateful to Professor Graham Shipley for drawing this find to my attention.
25 Michels 1967: 125.
26 See Hannah 1993a for a full discussion of the timekeeping clues in this passage.
27 See Taub 2003: 20–37 and Lehoux 2007 on these almanacs and their purpose.
28 Kidd 1974: 27–28.

3 MARKING TIME

1 By 'calendar' I assume something broader than a schedule of dates for years tied to a particular epoch, such as the Julian or Gregorian calendar. I allow for a 'lunar calendar', for instance, which may be tied to phases of the moon without any attachment to assigned dates in a year. In the scholarly literature on the *parapēgma*, the term 'calendar' has been used until recently without any misgiving, especially in the German tradition. I remain inclined to retain it. For a discussion of the different meanings and implications of the term, see Lehoux 2007: 70–75.
2 See Hannah 2005: 101–6, with a detailed explanation of the entries for the month of June from the *Fasti Antiates Maiores* (Figure 3.1, the sixth column from the left, near the centre). For a comparable situation in the Athenian calendar see Mikalson 1975: 186–97 (with apposite review by Lewis 1977), and Loraux 2002: 171–90.
3 Bromley 1986; Price 1974; Zeeman 1986.
4 See the website of the Antikythera Mechanism Research Project at: http://www.antikythera-mechanism.gr (accessed 2 September 2008).
5 The reconstructed solid models are referred to at: http://www.antikythera-mechanism.gr/data/models/solid-models.

The best images of a working reconstruction are currently of Michael Wright setting the dials to work on his life-size reconstruction in Jones 2008.

Some virtual reconstructions or animations, which are usually only partial, are available on the internet, but may be ephemeral, e.g. http://www.nature.com/nature/videoarchive/antikythera/ (accessed 1 August 2008) and http://www.youtube.com/watch?v=DMITkTYiZ7k (accessed 18 April 2008) and http://www.youtube.com/watch?v=qsr62p4h4Y8 (accessed 18 April 2008).
6 The digital radiographs of the 82 known remaining fragments of the Antikythera Mechanism are available for download from the website of Shaw Inspection Systems: http://www.shawinspectionsystems.com/library/antikythera/dr/dr.htm

The interactive Polynomial Texture Maps (PTMs), which result from the reflectance imaging, are available from the HP Labs website: http://www.hpl.hp.com/research/ptm/antikythera_mechanism/full_resolution_ptm.htm.
7 Bromley 1990; Freeth et al. 2006, 2008; Wright 2002, 2003a, 2003b, 2004, 2005a, 2005b, 2005c, 2005d, 2006a, 2006b. A popularised version of the story is available at Seabrook 2007.
8 Bromley 1986; Price 1974; Zeeman 1986.
9 Freeth et al. 2006, Freeth et al. 2008: 616.

10 Cf. the work of the following stone cutters identified by Tracy: the Cutter of *FD* III 2 no. 24, 138/7–128/7 BC (Tracy 1990: 170–72, Figure 26); the Cutter of *IG* II^2 937, ca. 135–123/2 BC (Tracy 1990: 173–80, Figure 27, Plates 25–26); and the following datable inscriptions on display in the Epigraphical Museum in Athens: EM 7751 (*IG* II^2 957), 158/7 BC; EM 7670 (*IG* II^2 1134), 117/6 BC; EM 7667 (*IG* II^2 1136), 106/5 BC. A major difference between any epigraphical parallel for the script and the writing on the Mechanism is that the Mechanism's inscriptions are extremely small in scale, but it is remarkable how they still share even the serifs.

11 For the distinctions see Ashmore 2007a: 91. I disagree with his definition of a sundial as a 'time finder and not a time keeper': through daylight hours, a sundial is undeniably a time keeper, a clock. Seely 1888: 28 seems to me to be more on the right track, regarding the sundial as the original time keeper.

12 Blackburn and Holford-Strevens 2003: 731–35; Hannah 2005: 12–15.

13 See Hannah 2005: 30–32.

14 Hannah 2005: 84; Parker and Dubberstein 1956: 1–2.

15 Hannah 2005: 43.

16 Hannah 2005: 32–41, 55–58.

17 Cf. Samuel 1972: 39–41.

18 Hannah 2005: 35–41.

19 In what follows I correct my response to this question at Hannah 2005: 55.

20 Hannah 2005: 30–31.

21 Davies 2003; Hannah 2001; Harris 1989; Thomas 1989.

22 Hannah 2005: 55–58, Samuel 1972: 42–49.

23 Toomer 2003: 196 states baldly that 'certainly the nineteen-year luni-solar cycle of Meton was derived from Babylon', and 2003: 970: 'The basis of the cycle (though not the year-length of $365^5/_{19}$ days) was undoubtedly derived from Babylonian practice.' But this is simply assertion. I am inclined to agree, but I know of no firm documentary evidence.

24 Evans 1998: 186; cf. Freeth *et al.* 2008. Supplementary Notes, 12–14 on the Antikythera Mechanism's system.

25 Lehoux 2005: 136–37; I was not aware of this article when I wrote at Hannah 2005: 53 that this fragment belonged to Miletos I; Lehoux reviews the evidence and finds no absolute evidence for such a linkage, although it remains a possibility. On the Miletos *parapēgmata* fragments in general see now Lehoux 2007: 154–57, 180–81, 223–26, 478–80.

26 22 June is a day too early for the actual solstice, which took place on 23 June. For a discussion of this issue and of the possible Babylonian input into determining the date of the solstice, see Bowen and Goldstein 1988; Hannah 2005: 53–54.

27 Cf. Hannah 2005: 57.

28 Price 1999: 172–73.

29 Other such calendars are known from the demes of Eleusis, Erkhia and Teithras, and from the Marathonian tetrapolis, and date to the fourth century BC: Dow 1968.

30 Aratos, *Phainomena* 733–35; cf. Hannah 2005: 27, 42–43, 71–72, Kidd 1997: 425–26, Samuel 1972: 57. For the inherent difficulties in constructing lunar calendars from observation of the lunar crescent, see Dunn 1999; for comparable Babylonian practice, see Stern 2008.

31 Cf. Dinsmoor 1931/1966: 421 for exceptions.

32 Cavanaugh 1996; Pritchett 2001: 8. Jones (2007: 165 n. 2) objects to the hypothesis that the Metonic cycle was adopted to regulate the Athenian civil

calendar on the grounds that this should have regulated the intercalary month itself. This seems to me an unnecessary imposition, at least in the early stages of the use of the cycle. The Babylonian version of the cycle appears to have taken some time to bed down the intercalary month: Parker and Dubberstein 1956: 2–9. Cf. Stern 2000: 168, who, while noting that 'most scholars are now of the view that it [the Babylonian 19-year cycle] was instituted at the beginning of the 5th century', also more cautiously acknowledges that 'This cycle, however, was still subject to minor adjustments during the 5th century and later, which suggests that it was not completely "fixed".'

33 John Morgan's work, demonstrating the regular use of the Metonic cycle in the fixing of the Athenian calendar in the Hellenistic period, remains largely unpublished, but is well regarded among historians of Hellenistic politics. All that is currently available in the public domain is an abstract of a conference paper, Morgan 1996; cf. Habicht 1999: v–vi.

34 Dinsmoor 1931/1966: 421.

35 This is on the basis of Osborne 2000 and 2003, where the traditional reliance on a secretary cycle is convincingly undermined in favour of the use of the Metonic cycle.

36 Müller 1991: 85–89. I do not find the evidence adduced by Müller entirely convincing, since he relies on only 35 epigraphically attested years in that period, which leaves a very large number of years unattested, particularly between the two terminal dates. While he also makes use of an argument from statistical probability to support his hypothesis, he does not seem to consider the possibility that the Metonic cycle as such might not have been in use at all at some stage in this long period. Müller 1994: 128–38 simply assumes the 1991 argument remains in force, and builds further upon it. Cf. also Jones 2007: 165 n. 2 for further doubts, although he seems unaware of Osborne's work.

37 Blackburn and Holford-Strevens 2003: 696 (Chinese New Year), 722–25 (Jewish Passover), 791–800, 862–67 (Christian Easter).

38 Poole 1998.

39 Pritchett 1947 and Pritchett and Neugebauer 1947 are the standard references for such a view; see also Pritchett 2001 and Pritchett and van der Waerden 1961. Literary commentators on the *Clouds* repeat it: e.g. Dover 1968: 177, to line 626, and, to a lesser degree, Sommerstein 1982: 193–94, who is, however, aware of the continuing, contemporary correlation between interest-days (the 'twenties') and the phase of the moon. Historians of science also follow suit: see Jones 2007: 165 n. 2; Samuel 1972: 58. But Pritchett and Neugebauer ascribe to Athens practices from elsewhere in the Greek world, or anachronistically retroject to fifth-century Athens practices from the much more volatile, and culturally distinct, period of the Diadochoi in the early Hellenistic period. Dunn 1999 presents a more measured view, arguing against any widespread practice of tampering, and demonstrating that where it did occur, it was as a result of the extreme pressures of wartime; cf. Pritchett 1999 for a riposte. Rogers 1924: 86–87 noted: 'She [the Moon] is alluding to the changes created by the introduction of the Metonic cycle some eight or nine years before the original exhibition of this Comedy . . . No doubt its introduction occasioned, at first, much the same disturbance as the adoption of the Gregorian calendar in the year A.D. 1792 occasioned among ourselves. And, in particular, as the Moon here complains, festivals would fall on different days from what they had formerly done, and from what they would still do in states which had not adopted the cycle.' Cf. Hannah 2005: 47–52; Merry 1879: 79–80, Murray 1933: 96.

40 This was the view of Dinsmoor 1931/1966: 421, and Rogers 1924: 86–87. See

also Dunn 1999. An inverse situation may have existed for the Jewish calendar by the fifth century BC, whereby an originally seasonal festival calendar became subject to lunar reckoning: the Jewish calendar was originally tied to agricultural activities, so that Passover, for instance, fell when the barley crop was ripe, and Pentecost when the wheat harvest was starting; but with the adoption of the Babylonian calendar the Jewish months and their festivals became tied to the moon, to the point that 'The festivals may thus have fallen out of line with their original agricultural datings': Stern 2000: 170–71.

41 Cf. Jones 2000: 156–57.

42 Bowen and Goldstein 1991.

43 On the planets and their character see Beck 2007: 70–90.

44 Grenfell and Hunt 1906, where the *parapēgma* is associated with a follower of Eudoxos; Lehoux 2007: 153–54, 217–23, questions how close a 'follower' of Eudoxos the compiler might really have been, but the matter is of no consequence here.

45 It is not clear whether the writer intended to mean the zodiacal constellations or their derivative signs of equal length of arc of 30°, but the former appears to be the more likely: cf. Bowen and Goldstein 1991: 246–48.

46 Cf. Grenfell and Hunt 1906: 152; Lehoux 2007: 217, 220.

47 On the Egyptian *lunar* calendar, see Hannah 2005: 85, with references.

48 On this and other early intercalary systems in Greece, see Hannah 2005: 29–41.

49 Griffiths 1970: 444.

50 Bennett (forthcoming), however, makes a compelling case that in the capital, Alexandria, a leap day was indeed inserted every fourth year to rectify the problem of the drifting calendar, but the change lasted only for a limited period.

51 Hannah 2005: 85–91.

52 Hannah 2005: 98–122, 131–35. See also Feeney 2007 on the cultural context of the adoption of the Julian calendar.

53 Bickerman 1980: 43.

54 Freeth et al. 2008: Supplementary Notes, 19–22.

55 See Feeney 2007: 138–66, esp. 139–42; Hannah 2005: 92–94, 146–57.

56 Bickerman 1980: 75–76; Feeney 2007: 18–19, 84–85; Samuel 1972: 189–94.

57 Price 1974. The term 'orrery' is probably best avoided, since it usually refers to three-dimensional models : see Taub 2006: 406–7.

58 Freeth *et al.* 2008: Supplementary Notes, 23. Omitted from the discussion here, but clearly important to the scientific significance of the Mechanism as a whole, are the eclipse dials. The principal lower dial is the Saros Dial, which presents an 18-year, 223-lunar month scale for predicting 38 lunar and 27 solar eclipses, indicated via 51 glyphs. The secondary lower dial is the Exeligmos Dial, which provides a triple Saros, 54-year cycle: Freeth et al. 2008: Supplementary Notes, 24–41.

59 Price 1974: 18. Most of these details are visible on the interactive PTM (AK31a) of fragment C, which is publicly available at: http://www.hpl.hp.com/research/ptm/antikythera_mechanism/full_resolution_ptm.htm.

60 Cf. Price 1974: 46. PTMs AK31a and AK34a provide the best available images at: http://www.hpl.hp.com/research/ptm/antikythera_mechanism/full_resolution_ptm.htm.

61 See Lehoux 2004, 2007: 66–69, and Taub 2003: 15–69 on the relationship between the *parapēgmata* and prediction. I tend to view *parapēgmata* from the perspective of the early devisers, for whom the observations of star phases may well have been direct not inherited, rather than of the later users, on whom Lehoux focuses. The astronomy of the *parapēgmata* still calls for analysis.

62 Price 1974: 46, 49.
63 Lehoux 2007: 187–88 also acknowledges this methodological leap of faith. An even more fundamental leap, one might add, is the assumption that the star list has anything at all to do with the Mechanism: it is not impossible that it was originally a separate item and has simply been wedded to the Mechanism by corrosive action in the sea.
64 Reiche 1989: 37–53.
65 Reiche, however, adds to Hesiod's star phases to make his point, thereby slightly undermining his argument.
66 Belmonte 2003; Hunger and Pingree 1989; Parker 1974.
67 Hannah 2002: 115–16, figs. 6.1, 6.2; Lehoux 2007: *passim*, and for the Milesian *parapēgmata*, 15 Fig. 1.3, 155–57; Taub 2003: 22–23 Figs. 2.1, 2.2.
68 For earlier literature on *parapēgmata*, see: Diels 1924/1965; Diels and Rehm 1904: 92–111; Ginzel 1911: 419–26; McCluskey 1998; Neugebauer 1975: 587–89; Rehm 1941; Rehm 1949; Schiaparelli 1926: 235–85; Taub 2003: 20–33; van der Waerden 1988: 76–79. On the impact of the mechanical clock, see especially Dohrn-van Rossum 1996.
69 Hannah 1993a, 1993b, 1997a, 1997b, 1998.
70 Lehoux 2000, Lehoux 2007: 154–57, 180–81, 223–26, 478–80.
71 The term also appears in literature: Cicero, *Letters to Atticus* 5.14, refers to an 'annual *parapēgma*'; this is of no assistance to understanding the name of the list in Geminos.
72 Hannah 2001b: 76–79, 2005: 59–61; Taub 2003: 20–27.
73 Diels and Rehm 1904: 104; cf. Lehoux 2005: 129–30, who sees no trace of any of the last line; see also Lehoux 2007: 180–81, 478–80.
74 Cf. Grenfell and Hunt 1906: 152, Lehoux 2007: 217, 220.
75 Cf. Lehoux 2005: 130–31, Lehoux 2007: 154–57, 223–26.
76 See Taub 2003: 26–37 on the function of these attributions.
77 This is not to say that this is exactly what Euktemon wrote, nor how he organised it. It simply demonstrates what we know of Euktemon's *parapēgma* from this source.
78 Hannah 2005: 59–70; Trevarthen 2000: 301.
79 Cf. Rehm 1913: 11–26, and most recently Lehoux 2007: 480–83 for the full text.
80 The attribution to Euktemon has been analysed in some detail by Lehoux, 2007: 181–87.
81 Blass 1887; Neugebauer 1975: 686–89.
82 Taub 2003: 15–69.
83 Price 1975.
84 Lehoux 2000, 2007; Rochberg 2004; Taub 2001, 2003.
85 See Lehoux 2007 for the most recent, in-depth discussion of Greek and Roman *parapēgmata*.
86 Lehoux 2000: 58.
87 Lehoux's conclusion was that: 'Since so many different possibilities are attested, and since most of the letters on the dial are insecure, I think any reconstruction of the entries for A through E, and Ω, would be doubtful. Moreover, there is no guarantee that all of the letters on the dial represented purely astronomical phenomena. They may have included important seasonal markers such as winds (compare Miletus I, above).' (Lehoux 2000: 59; see also 2007: 187–90). This last comment is an *argumentum e silentio*, which contrasts with the author's otherwise strong adherence to a positivistic methodology in dealing with the *parapēgmata*:

the Mechanism provides absolutely no evidence for meteorological observations, so introducing them from nowhere to detract from an argument is methodologically flawed.

88 Wright 2005b.
89 Hannah 2006b, 2007b.
90 Freeth et al. 2008.
91 Freeth et al 2008: Supplementary Notes, 15; cf. Trümpy 1997: 160–63 §132, updating Samuel 1972: 79.
92 Trümpy 1997: 155 §129.
93 Freeth et al. 2008: Supplementary Notes, 17.
94 Cabanes 2003: 88; cf. Trümpy 1997: 6, 159; Samuel 1972: 137.
95 Freeth *et al.* 2008: Supplementary Notes, 16.
96 Yunis 2001: 29–31, with further bibliography.
97 The equation of Macedonian Lōios with Athenian Boēdromiōn in the Demosthenic 'letter' presents a more fundamental problem. It cannot be reconciled with what we know from other sources, such as Plutarch (*Alexander* 3), who equates Lōios with Athenian Hekatombaiōn, two months earlier than Boēdromiōn. Yet Plutarch's identification dovetails perfectly with his further correspondence (at *Camillus* 19 and *Alexander* 16) between Macedonian Daisios (two months before Lōios) and Athenian Thargēliōn (two months before Hekatombaiōn), and both synchronisms agree with what we know of the later Macedonian calendar in its correlation with the Babylonian calendar, which would place Lōios around the summer solstice (Hekatombaiōn was the first month after the solstice). It would take more than a single intercalary month to have caused a shift of two months between the Macedonian and Athenian calendars to get Lōios to agree with Boēdromiōn, and it is as well to keep in mind that we know relatively little about the pre-Hellenistic state of the Macedonian calendar: cf. Hannah 2005: 82–85, 91–96.
98 Pritchett 1946.
99 Trümpy 1997: 142, 154, 178, 217, 244.
100 Trümpy 1997: 25–29; cf. Cabanes 2003: 93 n. 22 for doubts about Trümpy's interpretation of the meaning of the name Panāmos.
101 Freeth et al 2008: Supplementary Notes, 16.
102 Bennett 2004; Buxton and Hannah 2005: 302. See further Bennett (forthcoming) for strong arguments supporting the lunar alignment of the Macedonian calendar in Hellenistic Egypt.
103 Osborne 2003.
104 Freeth et al. 2008: Supplementary Notes, 12.
105 Cf. Hannah 2005: 47–52.
106 Gibbs 1976: 342–45, no. 5001; Hannah 2008: 753–54; Kienast 1993, 1997, 2005; Noble and Price 1968; Schaldach 2006: 60–83; Stuart and Revett 1762/2008: 12–25; von Freeden 1983. Schaldach 2006: 61–63 dates the Tower to ca. 100–90 BC, on astronomical rather than architectural grounds. For Gibbs (1976: 78–79) the Tower was dated to the first century BC.
107 Cf. Hannah 1997c for further discussion and references.
108 Seiradakis, pers. comm.
109 Neugebauer 1941–43: 209–50, cf. Neugebauer and Parker 1969: 225–35.
110 Beck 2006: 130–31.
111 Price 1975.
112 Cf. Taub 2002 for further exploration of the wider cultural uses of scientific instruments in antiquity.
113 Wright 2006b.

114 Beck 2006: 125.
115 For more on the ancient philosophies of time *per se*, see, for example, Sambursky and Pines 1971, Sorabji 1983 and Turetzky 1998.
116 Cf. Sambursky 1956: 59.
117 See further Pedersen and Hannah 2002.

4 TELLING TIME

1 Lippincott 1999: 108 no. 101; Symons 1998; Turner 1990: 59.
2 Bremner 1993; Diels 1924/1965: 157; Friberg *et al.* 1990: 498–99; Hunger and Pingree 1989: 96–101, 153–55 on MUL.APIN II.ii.21–42; Neugebauer 1975: 544–45; Schaldach 2006: 5–20; Symons 1998.
3 Compare Schaldach 2006: 5.
4 On the rejection of the attribution by Pliny of the same discovery to Anaximenes, see Schaldach 2006: 23–24 nn. 31–2. Pliny's testimony is: 'Anaximenes the Milesian, the follower of Anaximander, of whom we have spoken, was the first to display at Lacedaemon the horologium, which they call sciothericon' (Pliny, *Natural History* 2.187). Schaldach 2006: 4 also sees difficulties in the great gap in time between late sources like Diogenes and Pliny and the events they describe.
5 Proklos (*Commentaries on the First Book of Euclid's Elements*, p. 283, 7–10) says: 'Oinopides was the first to examine this problem, thinking it useful for astronomy. He, however, archaically calls the perpendicular "like a *gnōmōn*" (*kata gnōmona*), because the *gnōmōn* is also at right angles to the horizon'. Cf. Heath 1956: 271–72; cf. 181, 185, 370–72; Edwards 1984: 8–9.
6 Diels 1924/1965: 157; Gibbs 1976: 6–7. Thales' contemporary use of the human shadow to measure the height of the pyramids in Egypt seems a related activity: Diogenes Laertius 1.27 and Pliny, *Natural History* 36.82.
7 Gibbs 1976: 6.
8 Dicks 1970: 45; Dicks 1966: 32–3.
9 The precise astronomical location of the equinoxes does require a geometrical conception of the cosmos, as Ruggles 1999: 148–51 points out.
10 This is one of the usual methods available to prescientific societies; for a brief discussion, in the context of prehistoric henge monuments in Britain, see Trevarthen 2000: 301–2. Cf. the 'sun observation device' on Easter Island: Heyerdahl and Ferdon 1961: 228–29 with fig. 61. Similarly, some Mayan structures might have been oriented to allow for equinoctial shadow effects: Freidel *et al.* 1993: 34–36, 155–56. More controversially still, it has been argued that as early as the Neolithic period the Egyptians were aware of the equinoctial points, using the method described in the text here: see Sellers 1992: 28–32. What is interesting for our present purposes is that none of these cultures would be regarded as having had the theoretical framework which Dicks considered essential for the accurate placement of the equinoctial points.
11 Cf. Soubiran 1969: lix, on this 'inconvenient' usage.
12 Gibbs 1976: 7.
13 See Hannah 1989.
14 Gibbs 1976: 7, 275 no. 3055G (in Berlin, Pergamon Museum).
15 Lloyd 1988: 34–35.
16 Heath 1956: 370; cf. West 1973, who also (63 n. 2) makes the tentative suggestion that Herodotos' *gnōmōn* might have been a water clock; Proklos' testimony, however, regarding Oinopides' use of the term in the fifth century BC would seem to undermine this.

17 Euripides, *Orestes* 1685: West 1987: 176–77; Euripides, *Chrysippos* fr. 839.11 N: Diggle 1998: 166, Kannicht 2004: 877–81. Compare Dunbar 1995: 192, who refers also to *Prometheus Bound* 429 for a further instance of this usage for *polos*; this refers to Atlas carrying the 'heavenly dome' (*ouranion polon*) on his back, but the authenticity of *Prometheus Bound* 425–30 as a whole has been questioned: Griffith 1983: 161–62, West 1990: 425; the scholiast (Dindorf 1962: 21) states only that '*polos* is properly the tip of the axis' of the celestial sphere, suggesting he had problems with the sense implied in l. 429. *Polos* as the axial pole occurs in the *Peirithous* (fr. 594.5 N), attributed to either Kritias (Kannicht 2004: 606, Kannicht 1986: 173; compare Dillon and Gergel 2003: xiv) or Euripides (Mills 1997: 257–58). The reading of *polon* in the Euripidean fr. 911.3 N has been emended to the adjectival *poulun* (i.e. *polun*): Kannicht 2004: 918.

18 Robertson 1940; on the *hēliotropion*, cf. Bowen and Goldstein 1991: 72–77, Turner 1989: 310–11. In Aristophanes' *Birds* (l. 179) *polos* is used in a different sense, for the axis of the celestial vault or sphere: Dunbar 1995: 192.

19 See Cam 2001: 153 for a recent expression of this view; see also Soubiran 1969: 243, Edwards 1984: 12.

20 See Edwards 1984: 12–13 for the notion of a development through spherical to plane sundials. See Schaldach 2006: 3–4, 21 n. 8 for some modern views of the *polos* and its supposed place in the history of sundial technology.

21 Compare Schaldach 2006: 3–4.

22 On 'mid-quarter' days see Hannah 2005: 26, 64; Trevarthen 2000: 301.

23 Seely 1888: 44.

24 Powell 1940, followed by, for example, Edwards 1984: 14; Rose 1970, and Bowen and Goldstein 1991: 240 n.12 also appear to sympathise.

25 Langholf 1973: 383.

26 Schaldach 2006: 4, 23 n.27, 116–21, esp. 196–98; Schaldach 2004.

27 Schaldach 2006: 3–4; Lloyd 1988: 34–36; Langholf 1973; Robertson 1940.

28 See, for example, Aineias the Tactician 22.24–25; Pliny, *Natural History* 7.215; Vitruvius, *On Architecture* 9.8.7. Seely 1888: 45 entertained the notion that the Greeks may not have been aware even of the varying length of the day around 600 BC; his own audience sought to disabuse him of this extreme view (49–50).

29 Bowen and Goldstein 1991: 239–40.

30 See below in chapter 4 for the case of Aineias the Tactician's *klepsydra* in the mid-fourth century BC, which still betrays no awareness of seasonal hours. Both forms of hours existed in Babylonian shadow-measuring technology contemporaneously, as they did in Greek astronomical writing: Friberg *et al.* 1990: 498 n.7, Bowen and Goldstein 1991: 240, with further references; Schaldach 2006: 196–98.

31 Gibbs 1976: 6–7.

32 The letter-forms of the dedicatory and explanatory inscription on the sundial have been dated by Petrakos to ca. 350–300 BC. In a columnar inscription below the explanatory inscription a certain Theophilos is named – as a dedicator? – but part of the inscription – the demotic, according to Petrakos – has been erased above the ethnic, which makes Theophilos an Athenian. Petrakos suggested that this erasure occurred after 322 BC. This would make sense if by then Oropos was not in Athenian hands, and so the ethnic needed to be added, since as an Athenian Theophilos would be a foreigner at Oropos, and his demotic removed, since it had less relevance outside Athens. Certainly after 322 BC Oropos passed from Athenian control following Athens' defeat in the Lamian

War. This would provide a slightly tighter timeframe of 350–322 BC for the sundial, on the basis of which Schaldach (2006: 120–21) suggests identifying Theophilos with the Athenian archon of that name in 348/7 BC. This is tempting, but unprovable and problematic. The Theophilos named on the sundial is probably just one of the dedicators, so his political standing is not obviously relevant. But the situation is, I think, more complicated. If we examine the inscription in more detail (excellent images are available on the CD in Schaldach 2006), it appears that the letter-forms for 'Athenian' match those for the explanatory inscription above, whereas the letter-forms for 'Theophilos' are quite different: the epsilon, phi, omikron and sigma – half of the letters in the name – all differ from their equivalents in the explanatory inscription and in 'Athenian'. So it seems to me that what was erased was not a demotic (or even a patronymic, 'son of so-and-so'), but another proper name, which has been replaced by Theophilos' name above the erasure. Why this should have happened, I cannot say. In this case, the ethnic 'Athenian' would be an original feature of the dedication, and could therefore indicate a period when Oropos was not under Athenian control and when Athenians, as foreigners to Oropos, needed to be identified by their city-origin (I am grateful to Dr Sean Byrne for an enlightening discussion on these issues). My own sense of the style of the letter-forms is that a broader date between the mid-fourth and mid-third centuries BC is possible: securely dated decrees post-300 BC share the same letter-forms, e.g. Athens, EM 8099, *IG* II^2.1270 (298/7 BC), and EM 7375, *IG* II^2.780 (246/5 BC). In the period under discussion, Oropos was under Boiotian control until 338 BC, when it was given to Athens by Philip II. So one might wonder if the sundial dates to ca. 350–338 BC. After 322 BC Athenian control of Oropos has been suggested: the site seems to have passed to and fro between the Diadochoi through the late fourth century, and Athenian control of Oropos looks likely under Demetrios Poliorketes, at least from 304 BC until the battle of Ipsos in 301 BC, on the basis of the inscription *SEG* 3.117, which is dated by its archon to 303/2 BC, and which refers to Athenian interests in Oropos at that time (Walbank 1982: 181 n. 35; Hornblower 1991: 279; Dr Pat Wheatley, pers. comm.). So a date for the sundial between ca. 300 and 250 BC, on political and stylistic grounds, seems not impossible. The dial, of course, remains extraordinarily early in its sophistication.

33 Langholf 1973; Schaldach 2006: 23 n. 26.

34 Bilfinger 1886: 10–19.

35 Bilfinger 1886: 16; although he does not say so, Bilfinger would seem to have taken into account the need to correct for standardisation of time into local civil time, and the equation of time so as to produce local solar time as would be shown on a sundial. If we assume a relative height for the *gnōmōn-stoicheion* in the form of a person, we still do not know whether the shadow length was fixed throughout the year, or was to be adjusted according to the season. As we shall see, different shadow lengths for the same meal time are mentioned by other comedy writers, but whether this reflects differently sized shadow-casters or is to create slightly different jokes, is unclear.

36 Schaldach 1998: 24–25. This would require a solar altitude of 35° for a '7-foot' person to produce a '10-foot' shadow. The sun will not actually achieve this altitude between about 14 November and 7 February. On 29 January, for example, it culminates at noon at 33°, producing a shortest shadow of the day of just over 10 1/2 'feet' from a '7-foot' person, so a '10-foot' shadow is not possible in this period. On 31 March, the sun reaches 35° at about 3:10 pm local solar time.

37 E.g. down to 1/45 of an equinoctial hour (1.3 minutes in our terms) around 300 BC in *P. Hibeh* 27. How practical such minute subdivisions were is not clear. Ptolemy more pragmatically uses only halves, thirds, fourths, fifths and sixths as his small fractions: Toomer 1998: 7.
38 Indeed they lasted well into the medieval period: see, e.g., Lippincott 1999: 107.
39 Gibbs 1976: 94–95, n. 15.
40 Compare Bilfinger 1886: 16.
41 Bilfinger 1886: 16.
42 Edwards 1984: 15.
43 Neugebauer 1975: 736–46.
44 Cf. Neugebauer 1975: 737–39, who regarded the zodiacally organised version as the earlier, followed by the calendrical (Alexandrian and Julian). At the end of the second century BC Miletos was using a *parapēgma* organised into zodiacal months (MI, 109/8 BC: Lehoux 2007: 180–81, 478–80), which could later metamorphose fairly readily into the Julian calendar's 'solar' months (Hannah 2005: 132–33.)
45 Neugebauer 1975: 738.
46 Neugebauer 1975: 739–40.
47 Borchardt 1920: 27.
48 Hunt 1940–45: 41–42.
49 See Heslin 2007, with earlier bibliography.
50 Bremner 1993; Edwards 1984: 10–11, 15; Friberg *et al.* 1990: 498–99; Schaldach 2004: 4; Symons 1998.
51 Soubiran 1969: 116–18, 241 (Berosos), 257–58 (Patrokles); Edwards 1984: 10–12 (Patrokles).
52 See Dohrn-van Rossum 1996: 217–87, especially 236–45.
53 Turner 1990: 20.
54 McCluskey 2004: 199.
55 *Rule of the Master* 33.1–16; McCluskey 1998: 99, and generally 97–113.
56 Cf. Magnuson 2004 on the sometimes exaggerated modern views of the scale of Imperial and medieval Rome.
57 Gibbs 1976: 4; see Ardaillon 1900: 256 fig. 3883 for an imaginative illustration of the type.
58 Gibbs 1976: 45–46.
59 Compare Mills 1993 and Schaldach 2006: 43; Schaldach, 44–52 discusses the Byzantine examples from mainland Greece, and points out the apparent influence from Armenia.
60 Cf. Mills 1993: 84 on 'event markers'.
61 Istanbul, Archaeological Museum, inv. no. 905T: Schaldach 2006: 50, 56 n.67.
62 Mills 1993; note especially his computer-generated diagrams on p. 90 for seasonal-hour vertical dials at latitudes between 0° and 65°, and on p. 92 for the difference between true seasonal hours and equiangular hours on a hypothetical vertical dial for the latitude of Leicester (52.6°N).
63 That the zodiacal constellations have a Babylonian origin is demonstrated by the fact that most of the names used by the Greeks and Romans for them are simply translations of the Babylonian Akkadian names for these constellations: Koch-Westenholz 1995: 163–64, Bobrova and Militarev 1993.
64 Ardaillon 1900: 256–64 at 257; Rehm 1913: 2418–19; cf. also Hultsch 1909: 944–45, and Gibbs 1976: 60–61.
65 Schaldach 2006: 4, 23 n. 27, 116–21, esp. 196–98; Schaldach 2004.
66 Bowen and Goldstein 1991: 233–54.
67 Cf. the careful assessment by Soubiran (1969: 248–51) of the various

interpretations of the *arachnē*, concluding tentatively that it is a sundial whose receiving surface is patterned with a network of lines resembling a spider's web.

68 Veuve 1982: 23–36 (with further bibliography); other images of the sundial are available in R. Lane Fox, *The Search for Alexander* (London 1980) 430, and at http://en.wikipedia.org/wiki/Image:SunDialAiKhanoum.jpg (accessed 27 September 2007).

69 Hunt 1940–45: 41–42; noted by Gibbs 1976: 7, 94 n.12, but not included in her catalogue.

70 At midday 24 March the sun's altitude would be 51.6°, giving a shadow of 57.1 cm; at midday 24 April the sun's altitude would be 63.2°, producing a shadow of 36.4 cm.

71 At midday 23 March the sun's altitude would be 51.2°, giving a shadow of 57.9 cm; at midday 24 March, the sun's altitude would be 51.6°, providing a shadow of 57.1 cm.

72 At midday 23 March the sun's altitude would be 51.2°, giving a shadow of 120.6 cm; at midday 24 March the sun's altitude would be 51.6°, providing a shadow of 119.0 cm.

73 Gibbs 1976: 69, 158, no. 1044; Edwards 1984: 12.

74 Selçuk, Archaeological Museum, inv. 375: Gibbs 1976: 169–70, no. 1055G, pl. 10.

75 Roman sundials from Bulgaria are marked with lines denoting religious festival days and even the official birthday, *dies natalis*, of the emperor Antoninus Pius: Valev 2004: 55 n.3. I am grateful to Professor Tim Parkin for this reference.

76 Cf. Toomer 1998: 137 n.19 on the relationship with Aristarkhos.

77 Edwards 1984: 12.

78 Soubiran 1969: 247; cf. Ardaillon 1900: 257.

79 Edwards 1984: 12–13; Grenfell and Hunt 1906: 145. Stephanie West considers this particular *gnōmōn*, however, to be a water clock rather than a sundial, and Egyptian water clocks in general as the means by which Greeks learned to measure seasonal hours from the fourth century BC: West 1973; cf. Lewis 2000: 363. The precision of the time measurements recorded in the papyrus down to 1/45 of an equinoctial hour (i.e. 1 minute 20 seconds in our terms) might suggest a more sophisticated instrument than a sundial in this case, but the precision is illusory. Each day in *P. Hibeh 27*'s calendar is greater or less than the next by a regular 1/45 hour except for the five epagomenal days, when the sun, and with it time, is 'frozen'. In other words, the differences in day length are the product of a theoretical model of regular motion (barring the five days), of Egyptian origin, and not of actual observation, and in themselves tell us nothing of the instrument used: cf. Spalinger 1991: 356–58; and Dohrn-van Rossum 1996: 282–83 regarding 'ancient and theoretical but not measurable time-units' expressed in minutes and seconds in astronomical or astrological discussions.

80 Bergama, Archaeological Museum: Gibbs 1976: 274 no. 3054G, Plate 40.

81 Out of a total of 256 sundials in her catalogue, Gibbs (1976) lists 98 spherical, 109 conical, 40 plane, and 9 cylindrical. The total is greater now: Schaldach 1998: 40 puts the figure at 340, but does not specify the relative numbers of the various types.

82 Gibbs 1976: 62–63, 73, 268–69, no. 3049G; Gibbs 1979: 44 fig. 3. Edwards 1984: 10 disusses in detail the evidence for the date, which derives from a dedicatory inscription on this Asia Minor dial to a King Ptolemy – this is presumably one of the Ptolemies between Ptolemy III Euergetes, who took control of several cities in Asia Minor in 246–241 BC, and Ptolemy V

Epiphanes, who lost them in 203–201 BC (Gibbs 1976: 63 presents a different pair of Ptolemies, based on Rayet, but her version is mistaken) – and Vitruvius' attribution of the invention of this type of dial to a certain Dionysodoros, perhaps of Kaunos in Karia in Turkey, who was a contemporary of Apollonios of Perge, who worked ca. 200 BC.

83 Cf. Neugebauer 1948, who proposed a link between the actual discovery of conic sections ca. 350 BC by Menaichmos, a pupil of Eudoxos, and a type of sundial which he has difficulty exemplifying from the archaeological record, with only London BM 2546 (= Gibbs 1976: 363–65 no. 5022G, pl. 59) presenting itself for comparison. Cf. Gibbs 1976: 62.

84 Cam 2001: 160–62, fig. 12, for a template for making a conical sundial, based on Rayet 1875; Gibbs 1976: 17, 74–75, 77; Mills 2000b: 64.

85 Gibbs 1976: 78, 324–25, no. 4001G, pl. 52. Cf. Edwards 1984: 10, who mistakenly reported that Gibbs dated this dial to the third century BC.

86 See chapter 3. For the surviving roofed spherical dial designed by Andronikos and now in Tenos (Archaeological Museum, no. A139), see Gibbs 1976: 71, 373–75 no. 7001G. On Andronikos, see Müller 2001.

87 See Schaldach 2006: 68–83, Gibbs 1976: 342–44 no. 5001 and Delambre 1817: 487–503 for explanations of the mathematics of these sundials.

88 Schaldach 2006: 116–21 no. 23.

89 E.g. Soubiran 1969: 257–60; Edwards 1984: 10. Pattenden (1979: 203–6, and 1981) accepts the identification, but does not discuss the attribution.

90 Cam 2001: 152–55, with 155 fig. 8. It must be admitted that the *pelecinum* exists in Faventinus' text only by emendation from the manuscripts' unanimous *pelignum*.

91 Cf. Gibbs 1976: 61.

92 Two examples from Akradina, perhaps no more than 10 cm in height, and one from Neapolis, only about 5 cm high, are on display in the Museo Archeologico 'Paolo Orsi', in Syracuse, Sicily.

93 London, British Museum EA 68475; Evans and Marée 2008.

94 Evans and Marée 2008: 13.

95 Price 1969: 244–46, Schaldach 1998: 41–42.

96 Arnaldi and Schaldach 1997.

97 Arnaldi and Schaldach 1997; Hannah 2008: 751–52; Price 1969.

98 Oxford, Museum for the History of Science, inv. no. 51358; Price 1969: 253–56; Schaldach 1998: 45–47; cf. a Byzantine example, dated ca. AD 400–600: London, Science Museum inv. 1985–222, in Oleson 2008: 794 figure 31.3.

99 Istanbul, Archaeological Museum, inv. 2970.

100 Ashmore 2007a, 2007b, 2007c, 2008; Evans 1999: 253–54; Neugebauer 1975: 868–79; Price 1957: 603–9; van Cleempoel 2005; Taub 2001: 920–21.

101 Edwards 1984: 10–12.

102 Gibbs 1976: 64.

103 Schaldach 2006: 31.

104 Cf. Turner 1994: 29–31, 34–36.

105 Oropos: Schaldach 2006: 116–21 no. 23; cf. the sundials from Ai Khanoum: Veuve 1982: 36–51.

5 MEASURING TIME

1 Athens: Gibbs 1976: 74, 227–28 no. 3008G, pl. 28., Schaldach 2006: 91–93 no. 2, 184; Klaros: Gibbs 1976: 270 no. 3015G, Martin 1965: pl. XV.4.

2 I am grateful to Professor John Oleson for this reference.
3 Modern experiments on replica lamps suggest that they could burn for several hours: cf. http://www.landesmuseum-fuer-vorgeschichte-halle.de/lightkultur/luxluxus/index.htm (accessed 28 November 2007) for a scientific discussion of the lighting capacity of ancient lamps.
4 Preisendanz 1973: 1–16, 124–34, 303–6; Betz 1986: 172–82, and 336 on 'Lamps, not painted red'. On lamp magic, see Eitrem 1991: 176–79; Mastrocinque 2007.
5 Betz 1986: 10–12; Preisendanz 1973: 14–18.
6 Betz 1986: 175; Preisendanz 1973: 93.
7 [Bishop Asser], *Life of King Alfred* 103–104, see Smyth 2002: 51–52, 244 and Stevenson 1959: 89–90, 338–41. Smyth argues for a date of composition ca. AD 1000, about a century after Alfred's and Asser's lifetimes, and debates the authenticity of stories like these attached to Alfred.
8 Dohrn-van Rossum 1996, *passim*; Turner 1993: 161–72.
9 Schaldach 2006: 4, 23 n.27, 116–21, esp. 196–98; Schaldach 2004.
10 Cotterell *et al.* 1986.
11 Lewis 2000: 363; West 1973.
12 Dohrn-van Rossum 2003: 462; Lewis 2000: 344, fig. 1.
13 Cf. Stocks 1930: n.1 at *De Caelo* 294b22; Diels 1924/1965: 192, Abb. 66; Pattenden 1987.
14 Lewis 2000: 344.
15 Dohrn-van Rossum 2003; Humphrey 2006: 101–103; Lewis 2000: 361–69; Theodossiou and Kalyva 2002.
16 Young 1939.
17 Young 1939: 281; the figure might vary, as no allowance was made for variations in the rate of flow owing to variations in the temperature or quality of the water.
18 Cf. Seely 1888: 38.
19 Cf. Pattenden 1987: 168.
20 Seely 1888: 36–37.
21 Young 1939: 276 n.7.
22 Allen 1996: 159.
23 Fornara 1983: 102–104 no. 97; Meiggs and Lewis 1988: 111–17 no. 45.
24 Hicks 1887: 107–11; Young 1939: 278 n.27.
25 Young 1939: 281 n.41.
26 At the time of the midwinter solstice the length of daylight in Athens is 9 hours 25 minutes. Young 1939: 281 n.41 more loosely reports a range between 9 hours 28 minutes and 9 hours 34 minutes for December-January, which corresponds to the range between mid-December and mid-January.
27 Diels 1924/1965: 198–99 with Abb. 68; cf. Bilfinger 1886: 9–10.
28 Lewis 2000: 363; Soubiran 1969: 70–71.
29 Cf. Bowen and Todd 2004: 110.
30 Toomer 1998: 252 n.51. On the date of Kleomedes, see Bowen and Todd 2004: 2–4.
31 Evans and Berggren 2006: 163.
32 Seely 1888: 43 used an alternative reading of Pliny, which gave him twenty water clocks plus four, to reduce the lawyer's speech to one of just an hour and a half.
33 Crawford 1996: 393–445; see also Hardy 1912: 7–9, 46–47. The tablets of the law are now in the Museo Arqueológico Nacional, Madrid.
34 On Roman proceedings in law courts see Jones 1972: 69–73, and Greenidge 1901/1971: 456–504, esp. 459, 476–77; I am grateful to Dr Jon Hall for this last reference and the following one, and for discussions about Roman legal practice in general.

35 Lintott 2004: 77.
36 Hall 2006.
37 Other types of cases in other jurisdictions in the Roman world attracted different regulations: see Lintott 2004.
38 Harvey 2004: 180–81 no.159. I am grateful to Dr Jon Hall for this reference.
39 Cf. Feeney 2007: 213–15.
40 The story would seem to suggest that the inaccuracy among sundials was not a matter of one dial telling one hour and another an entirely different hour, but of telling at what part of a given hour an event took place.
41 Cotterell *et al.* 1986: 39.
42 Dohrn-van Rossum 1996: 323–24.
43 Dohrn-van Rossum 1996: 324–25 mentions various examples of what he calls 'clock telegraphy', but warns against presuming that this ancient technology was a direct precursor to nineteenth-century telegraphy.
44 Cf. Whitehead 2002: 111–13.
45 Borchardt 1920: 6–26; Cotterell *et al.* 1986.
46 Lewis 2000: 363; West 1973.
47 Young 1939: 277.
48 West 1973: 63.
49 Pattenden 1987.
50 Lewis 2000: 362; Whitehead 2002: 158–60.
51 Armstrong and Camp 1977; Camp 1992: 157–59.
52 Petrakos 1968: 113–16; Turner 1990: 62–63, no. 65. Other water clocks have been found at Samos, Priene and Pergamon: Armstrong and Camp 1977: 151 n.3; Dohrn-van Rossum 2003: 463.
53 Cf. Armstrong and Camp 1977: 155 fig. 5 for a hypothetical reconstruction.
54 Humphrey 2006: 102–3.
55 Armstrong and Camp 1977: 154 n.9; Borchardt 1920: 6–7, Taf. 1–2; Lewis 2000: 361–62, Fig. 8; Neugebauer and Parker 1969: III pl. 2; Turner 1990: 58, no. 49; von Bomhard 1999: 12–13, 15–17 Figs. 11–12.
56 Armstrong and Camp 1977: 154.
57 Humphrey 2006: 103, with 101 Fig. 18; Lewis 2000: 363–66.
58 Armstrong and Camp 1977: 157–58.
59 For reconstructions see Dohrn-van Rossum 2003: 462; Humphrey 2006: 102 Fig. 19; Lewis 2000: 364 Fig. 9.
60 Lewis 2000: 364–65.
61 Dohrn-van Rossum 1996: 26–27; Drachmann 1954; Lewis 2000: 366; Price 1957: 601–3; Rowland and Howe 1999: 117, 290–91; Stierlin 1986: 235–45. For a broader context in the realm of religion, see Beck 2006: 200–6.
62 The translation is that of Joseph Gwilt, *The Architecture of Marcus Vitruvius Pollio* (London: Priestley and Weale, 1826), provided online, with commentary, at Bill Thayer's website, *LacusCurtius: Into the Roman World*: http://penelope.uchicago.edu/Thayer/E/Roman/Texts/Vitruvius/home.html. The text is reproduced here by kind permission of Bill Thayer.
63 Abry 1993; Bendorff *et al.* 1903; Evans 1999: 251–53; Gundel 1992: 208–9 no. 17; Künzl 2005: 88–91; Lewis 2000: 366–67; Noble and Price 1968: 352; Turner 1993: 95–110.
64 See chapter 3.
65 Stuart and Revett 1762/2008: 15–16; see also Salmon's comments in the new Princeton edition, xiii.
66 Noble and Price 1968; Kienast personal communication.
67 Price 1975: 367.

68 Price 1975: 368.
69 Price 1975: 368.
70 Price 1975: 369.
71 Gibbs 1976: 86.
72 Remijsen 2007. See further chapter 6.
73 Edwards 1984: 10; Gibbs 1976: 62–63, 73, 268–69, no. 3049G.
74 Edwards 1984: 12; Gibbs 1976: 69, 158, no. 1044.
75 Langholf 1973. Edwards (1984: 14) assigned the invention of the division of the day into twelve *hōrai* to the time of Pytheas of Messalia (fourth century BC) and Timokharis (third century BC); earlier instances, such as Herodotos 2.109, he regarded as dubious and possible interpolations (but on this see now Lloyd 1988: 34–36). The reference from Pytheas occurs in a quotation by Geminos (6.9). The difficulty is whether the word *hōrai* lies inside the quotation, and therefore was used by Pytheas, or not: cf. Roseman 1994: 140, 143, who accepts Mette's text of Geminos, which attributes the use of *hōrai* to Geminos; whereas Evans and Berggren 2006: 162, accept Aujac's emendation of Geminos, which attributes the use of *hōrai* to Pytheas himself.
76 Edwards 1984: 14.
77 Schaldach 2006: 116–21, 196–98 no. 23. See also West 1973, who argued for a date of inception for the equinoctial hour in Greece, adopted from Egypt, about 300 BC, on the basis of *P. Hibeh 27*; Bowen and Goldstein argue similarly for both the equinoctial and seasonal hours; none of these appears to have been aware of the Oropos sundial. See above, chapter 4.
78 Cf. Edwards 1984: 14.
79 Edwards 1984: 14.
80 Seely 1888: 46: 'It was not until Europe had emancipated herself from slavery to this most awkward of time systems that modern time-keeping became possible. For many centuries invention was, as it were, thrown off the scent by the necessity of converting the regular and uniform motions which could be given to mechanism into means for displaying the ever-varying hours of the Roman system.' (By 'Roman system' Seely means the use of seasonal hours, which he mistakenly attributes to the Romans.) Cf. Humphrey 2006: 103.

6 CONCEPTIONS OF TIME

1 The incomplete nature of Vitruvius' instructions leaves obscure what type of sundial he had in mind for it; Veuve 1982: 31 thought the *analemma* was intended for a spherical sundial, whereas Soubiran 1969: 233 thought it was meant for a horizontal plane dial. Much is owed to Bilfinger 1886: 27–37, where the *analemma* is applied to the construction of a spherical and a horizontal sundial; Drecker 1925: 3–4 applied it to horizontal and vertical plane dials and to a cylindrical dial; see also Evans 1999: 247–51, Gibbs 1976: 105–9, Rowland and Howe 1999: 289, and Soubiran 1969: 234–40.
2 Evans 1999: 249; Evans and Berggren 2006: 27–34.
3 On Eudoxos' use of a globe, see Evans 1999: 239. Hipparkhos' globe is mentioned by Ptolemy, *Almagest* 7.1 (H12).
4 Naples, Museo Archeologico Nazionale inv. 6374; Evans and Berggren 2006: 28–29, Fig. I.2; Gundel 1992: 204, 207 no. 8, and endpapers; Künzl 2005: 60–77, esp. 63–65, Abb. 6,3, 6,4, 6,7, 6,8, 6,10; Schaefer 2005; Stott 1991: 6. A much smaller globe (11 cm in diameter, in contrast to the Farnese Atlas' 65 cm diameter globe) is a second-century AD bronze specimen in Mainz, Römisch-Germanisches Zentralmuseum inv. O.41339: Evans and Berggren

2006: 28, 30–31 figs. I.3–4; Künzl 2005: 60–63, Abb. 6,1, 6,2, 6,6, 6,9, 6,11, 6,12.

5 Evans and Berggren 2006: 32–33, fig. I.5; von Boeselager 1983: 56–60, pl. XV fig. 29–30.

6 It is assumed that it belongs to the same period as the house's First Style wall decoration; although it is an *emblema* set within the larger floor, it appears not to have been prefabricated on a separate base (possibly at an earlier date) and imported, but rather to have been laid *in situ*: von Boeselager 1983: 57, 60; Westgate 2000: 261, 264 n. 39.

7 Cf. von Boeselager 1983: 57, who thinks the ring which I have called the equinoctial colure cannot be that, since it does not cut through *both* of the points of intersection between the equator and the ecliptic, but is a meridian. It is true that this ring does not cut through both the intersection points, but by the same token it cannot be some other meridian, because as another great circle, if a meridian cuts one intersection point of the equator and ecliptic (as this ring does), then it must cut the other, and can be none other than the equinoctial colure. In my view, the artist has simply lost track of where the ring should cut the other rings and has made a mistake. Even Boeselager acknowledges that as a meridian, this ring is 'willkürlich eingezeichnet', so that it is impossible to tell what location on earth the supposed meridian is meant to signify. So why can it not be an 'arbitrarily drawn' colure?

8 Cf. Rudolf Schmidt's remarks that the outer meridian ring would allow 'adjustment of the instrument to the Celestial Pole (polar height) for the performance of astronomic calculations (such as establishment of day lengths for a particular location, for every calendar day; or determination of the geographic locations of sunrise and sunset, or calculation of maximum sun heights at noon)'. (http://www.coronelli.org/publikationen/news/2003/news2003_e.html: accessed 1 November 2007). This seems to me to be an over-optimistic interpretation of a poorly preserved part of the mosaic.

9 von Boeselager 1983: 60. A later Roman tomb in Syracuse is still popularly called 'Archimedes' Tomb'. A later painting from Stabiae near Pompeii, dating before the eruption of Vesuvius in AD 79, illustrates another armillary sphere, but is less well preserved: Arnaud 1984: 73; Picard 1970: 84 pl. LVIII (colour).

10 Pedersen and Hannah 2002: 65–66. Cf. Evans 1999: 241.

11 Vitruvius' numbering system is confused in the surviving manuscript tradition, he does not use degree notation, nor trigonometric functions. My presentation of the *analemma* seeks to simplify a complex situation, and to express it in modern terms, in the interests of emphasising the interconnectedness between the designs of the spherical and conical sundial types.

12 This matches examples in Gibbs 1979, but there appears to be no mathematical underlying principle: perhaps it depended on external factors, such as the size of the block of stone or the desired sculptural form of the sundial (cf. 16 below, in the main text, for a possible internal geometrical principle).

13 This disproportion matches all but one example in Gibbs 1979.

14 Gibbs, 1976: 4, has 256; Schaldach 1998: 40.

15 Veuve 1982: 26–27, 36.

16 Piraeus Museum inv. no. 235: Schaldach 2006: 116–21, 196–98, no. 23.

17 Schaldach 2004: 440.

18 Schaldach 2004: 443.

19 For a summary of eastern dial forms and techniques, see now Schaldach 2006: 5–20.

20 At this point I agree with Dicks' assessment of the intellectual capacity of the Greeks (Dicks 1966: 32–33, 1970: 45).

21 Schaldach 2006: 196–98. The identification of the hours as preferably equinoctial rather than seasonal is hypothetical: the southern, 'winter' face of the dial certainly presents equinoctial hours, but the northern, 'summer' face appears to present seasonal hours. This latter face, however, is less carefully inscribed, and the explanatory inscription on the 'winter' face says nothing about the dial utilising two types of hours. Schaldach therefore assumes that a single type of hour was intended for the whole monument, and that this was meant to be the equinoctial hour, which has been inaccurately rendered on the 'summer' face. The counter-argument that the 'winter' face's hours have been inaccurately rendered, and present accidentally as equinoctial hours, seems improbable.
22 Schaldach 2006: 4 and n. 27.
23 Grenfell and Hunt 1906.
24 Eliot 1976: 656; Walbank 1982: 181 n. 36.
25 Gibbs 1976: 74, 227–28 no. 3008G, pl. 28; Schaldach 2006: 91–93 no. 2, 184; Stuart and Revett 1762/2008: 29, 33, Plate I.
26 Gibbs 1976: 226 no. 3007, 239 no. 3020; 1979: 45, fig. 4.
27 Pattenden 1979: 204, 207–8. Cf. Cam 2001: 157 n. 12.
28 Hüttig 2000: 173–75, regarding a sundial made for latitude 30°N (probably Alexandria, latitude 31°13′N), but transferred and used at latitude 40° (on the island of Kephalos). Cf. Schaldach 2006: 136–37, 207–8 no. 38, who disputes aspects of Hüttig's technical analysis of this dial. See also Pattenden 1981 for a sundial made for about latitude 31°N (probably Alexandria again), but used at latitude 37°40′N (Aphrodisias), again with tolerable accuracy despite the shift. Pliny, *Natural History* 2.182 was aware of the geographical limitations of sundials.
29 Houston 1989: 66 n. 17.
30 See the text and discussion of this passage in White 1984: 214; Houston 1989: 64 n. 8.
31 Houston 1989: 64.
32 Houston 1989: 64.
33 For scientific usage, see e.g. Dioscorides Pedanius (1.33.1, 1.204.2); Hipparkhos (*Commentary on Aratos and Eudoxos* 3.2.1.11, 3.2.4.3, 3.2.5.5, 3.2.6.11, 3.2.6.14, 3.3.1.6, 3.3.2.12, 3.3.4.9, *Geographical Fragments* no. 47); Strabo (2.5.36). Remijsen 2007: 129 restricts the use of the hour to religious and scientific contexts. On the Last Day and the equinox: McClusky 2007.
34 Gibbs 1976: 78, 324–25, no. 4001G, pl. 52.
35 Gibbs 1976: 342–45, no. 5001; Hannah 2008: 753–54; Noble and Price 1968; Schaldach 2006: 60–83; von Freeden 1983.
36 Delambre 1817: 487–503; Kienast 1993, 1997, 2005, and personal communication.
37 Gibbs 1976: 74, 227–28 no. 3008G, pl. 28; Schaldach 2006: 91–93 no. 2, 184; Stuart and Revett 1762/2008: 29, 33, Plate I. Other sundials from the Theatre of Dionysos are: Athens, National Museum 3157, 3158, and 3159 (Gibbs 1976: 224 no. 3005G, 220–21 no. 3001G, pl. 26; and Locher 1989). On the early history of the discovery of sundials, beginning in the Italian Renaissance, see Turner 1993: 208.
38 Armstrong and Camp 1977; Turner 1990: 62–63, no. 65 (cf. Aristotle, *Poetics* 1451a7–9); Turner 1994: 32–33. Cf. Soubiran 1969: 239 on the use of *klepsydrai* to check the accuracy of a sundial's gradations.
39 Cf. Gibbs 1976: 17, 74–75, 77; Schaldach 2006: 35–36.
40 London, British Museum 1936.3–9.1: Gibbs 1976: 304–5 no. 3086G, pl. 48; Mills 2000a: 9, 2000b: 66.

41 For the graphical method, cf. Valev 2004 (I am grateful to Professor Tim Parkin for drawing this article to my attention); for the mathematical, cf. Gibbs 1976: 4–5, 77.
42 Gibbs 1976: 75.
43 As Gatty noted, 'the inclination of the face of the block appears to correspond with the latitude of Alexandria' (Gatty 1890: 388). Cf. Gibbs: 'On a dial constructed with the aid of an analemma, the best measure of the latitude intended by the dial maker is the angle between the top and front surface; the next best indication of the intended latitude is provided by the relative distances on the meridian line; measurements taken on day curves give the least reliable indication of the intended latitude.' (Gibbs 1976: 77, cf. 33–35, 74–75).
44 Cf. Valev 2004: 55 n.3 regarding the commemoration of an emperor's birthday on Roman sundials.
45 Cf. Mills 2000a: 9, 2000b: 65, who found a 'triad' of holes on the seasonal lines which he believed served as markers for the hour lines; triplets of holes were not obvious to me when I studied the dial in 2005.
46 Gibbs 1976: 304. This latitude is disconcertingly equal to that of London (51°). One might suspect the 'solstitial' lines were the product of modern additions, if it were not for the 'probably Byzantine' (Gibbs 1976: 304) style of the Greek letters indicating the hours; cf. the table in Schaldach 2006: 221, which confirms the Byzantine character of the letters.
47 Evans and Marée 2007: 3; Gatty 1890: 388.
48 Note that Mills 2000b: 66 fig. 14 defines ω in degrees – 37° – for this sundial. A curious feature of this sundial, already noted, is the close similarity between the distance from the upper, 'winter solstice' line and the equinox, with the distance between the equinox and the lower, 'summer solstice' line. This occurs on only one other conical sundial in Gibbs' catalogue; all the rest show the expected difference between a shorter winter-to-equinox and longer equinox-to-summer meridian distance. Were this a spherical sundial, this is very much what we would expect, whereas a south-facing conical dial will make the upper distance usually much smaller than the lower (compare the *analemma* above). If, however, we were to construct the conical sundial with the base of its cone parallel not to the equinoctial line, but instead to the summer solstice line in the *analemma* (this is *not* the tropic itself, though), then the solstitial day curves would be almost correctly situated where they are. In that case, the apex of the cone would not be directed towards the North Celestial Pole.
49 Symons 1998: 30–31.
50 Buchner 1976, 1980, 1982, 1993–94; Hannah 2005: 129–30; see now Heslin 2007 for a full critique on why it appears unlikely that this was a full sundial.
51 Veuve 1982: 23–25.
52 Cf. Shelton 1998: 353–54.
53 Stuart and Revett 1762/2008: 16.
54 Akurgal 1993: 139; Gibbs 1976: 270 no. 3015G; Martin 1965: pl. XV.4. This is one of the largest surviving conical dials, with a width of 1.10 m (misprinted in Gibbs as 110 mm) and height of 78.85 cm, second only to the one which overlooks the Theatre of Dionysos in Athens (Figure 0.1). It is no longer *in situ*, and appears to have been moved off-site completely, but its present whereabouts have proved impossible to discover. I am grateful to Professor Juliette de La Genière for information on this sundial and related finds from Klaros.
55 Veuve 1982: 24 and n. 4.
56 A plaque on the column records a donation at their own expense by the *duoviri*, L. Sepunius and M. Herennius. Old images are reported to show a sundial

topping the column and it is assumed that this was the object of the donation (La Rocca *et al.* 1994: 107). The sundial currently on the column is a cast of one from Pompeii (Gibbs 1976: 287–88 no. 3066G). A similar inscription was found near the 'Temple of Hercules', i.e. the unidentified Doric Temple in the Foro Triangolare; it recorded the donation by the same two magistrates of another sundial (De Vos and De Vos 1982: 32; Gibbs 1976: 394 no. 8007). The sundial beside the Temple of Apollo may appear to work on a sunny day, but because it is set off-axis to the meridian and instead is aligned to the south-east with the temple, then in midwinter it will be about three hours slow, and in midsummer, when European Summer Time is in force, about two hours slow.

57 The identification of Apollo with Helios is first attested in literature in Euripides, *Phaethon* fr. 781. 11–12 (Kannicht: 2004: 817).

58 *CIL* 12.2522, *ILS* 5624; Harvey 2004: 91 no. 58; I am grateful to Dr Jon Hall for this reference.

59 Modern instances include the erection of public clocks in Turkey in the early twentieth century, as the new state moved from seasonal to equatorial hours.

60 Gibbs 1976: 90–92.

61 Cf. Seely 1888: 42. I am indebted to Professor John Barsby for the observation that Roman comedy generally did not satirise Greek customs.

62 Schilling 1977: 261 at §213.1.

63 One may wonder if the source of the sundial was the Samnite town of Aquilonia in Campania, sacked in 293 BC by Papirius and his fellow-consul, Spurius Carvilius Maximus. The temple of Quirinus had been vowed by Papirius' father, but was built by the son and decorated with spoils from the latter's victory (Livy 10.41–42, 46; Orlin 2002: 122–23, 135–36, 180. Oakley 2005: 449–50 refers to the passage from Pliny, but without linking the sundial to booty). Campanian Samnites had long been culturally hellenised (see Frederiksen 1984), so it is not impossible that they had adopted Greek timekeeping technology at an early stage (contrast Seely 1888: 40, who took the Samnites to be 'a ruder people even than the Romans' and therefore incapable of providing the Romans with a sundial).

64 Cf. Schaldach 2006: 33; Sefrin-Weis 2007: 33.

65 Gibbs 1976: 96 n. 25, cf. Hüttig 2000: 173–75. In midsummer there is about half an hour's (28.8 mins) more sunlight in the day in Rome than in Catania, but about half an hour's (27 mins) less sunlight in midwinter. We have to spread this over the twelve hours of the day (2.4 mins in summer, 2.25 mins in winter). So the seasonal hours were very marginally different in length between the two cities at these extremes.

66 Cf. Hüttig 2000.

67 Gibbs 1976: 91–92. But note in general the caveat in Hüttig 2000: 164, 167 regarding errors in Gibbs' calculations.

68 Gibbs 1976: 91, 329–30 no. 4005G, pl. 54.

69 See Gibbs 1976: 300, no. 3080 for a list of the sundials so marked.

70 See Dohrn-van Rossum 1996: 29–39. In the Eastern Orthodox tradition, the Book of Hours is still called a *Hōrologion*.

71 Cf. Beard *et al.* 1998: 143. Contrast Houston 1989: 65, who found no interest in scheduling among the Romans.

72 I am grateful to Dr Jon Hall for drawing my attention to this passage.

73 Cf. Dohrn-van Rossum 1996: 17, Schaldach 2006: 35.

74 Cf. Richardson 2004: 29.

75 http://www.spartathlon.gr (accessed 20 December 2007). I am grateful to Professor Keith Rutter for drawing this ultra-long distance race to my attention.

76 Ramsay 1920: 81.
77 Llewelyn 1994: 5 n.19.
78 For Cicero see Nicholson 1994; for private letters in general see Llewelyn 1994: 26–57.
79 Grenfell and Hunt 1906: 286–94; Llewelyn 1994: 6–7.
80 Remijsen 2007: 133.
81 Llewelyn 1993, 1994: 8–13.
82 Remijsen 2007: 134–35.
83 This system is often called the *cursus publicus*, but this term should be reserved for the civil equivalent of the military service designed by Augustus: see Remijsen 2007: 130, 135.
84 Casson 1994: 182–89; Llewelyn 1994: 1–25; Mitchell 1976; Ramsay 1920, 1925. It was suggested by W. M. Ramsay that the seven early Christian Churches of *Revelation* were chosen as recipients of John's letters because they were on the Roman postal route in that area (Ramsay 1904: 185–96), but evidence in support of this hypothesis still seems to be lacking: see Trebilco 2006: 25 (I am grateful to Professor Paul Trebilco for drawing this instance to my attention).
85 O. Krok. 83: Remijsen 2007: 138.
86 Dohrn-van Rossum 1996: 330–35.
87 Ramsay 1925.
88 Pliny, *Natural History* 2.180 (Sicily), cf. Ptolemy, *Almagest* 1.4, 2.1, Neugebauer 1975: 667–68, 938, Toomer 1998: 75 n.3; as Bowen and Todd 2004: 66–67 emphasise, lunar eclipses are more secure for this purpose, since solar ones are not visible everywhere.
89 Dohrn-van Rossum 1996: 323–50; cf. Braudel 1979: 424–28, especially the isochronic maps on pp. 426–27, indicating the slow speed of news from around Europe to Venice between 1500 and 1765.
90 On the 'Peutinger Table', the 'Vicarello cups' and the 'Antonine Itinerary' see Dilke 1985: 113–20, 122–24, 125–28. I am grateful to Professor Richard Talbert for advice on the date of the cups.
91 Arnaud 2007.

7 EPILOGUE

1 Gibbs 1976: 23–27, 194–218; Pattenden 1979.
2 On the date of the construction of the Pantheon, see Hetland 2007. De Fine Licht 1968 remains the fundamental study of the building. See more recently: Dumser 2002; Gros 1996: 173–80; La Rocca 1999; Stamper 2005: 126–29, 184–205; Virgili 1999; Ziolkowski 1999.
3 I became aware of this movement of the sun within the Pantheon before I discovered Sperling 1999, who elaborates on the idea in much greater detail.
4 While Grainger 1933 was sympathetic to the idea of the Pantheon as a sundial, he dismissed it as useless in Rome, and instead calculated it as relevant for Syracuse and Rhodes, seemingly unaware of the better parallelism between the building and the roofed spherical sundial, which would tie the building directly to the latitude of Rome.
5 Sperling 1999: 112–13 Abb. 59–60.
6 Voisin 1987 develops more fully the astronomy of the Golden House and the ideas underlying its usage. See also Oudet 1992, and Wilson Jones 2000: 24. On the Golden House and the Pantheon as 'images of the universe', see Beck 2006: 120–23, Stierlin 1986: 123–72.

7 Cf. Stierlin 1986: 25–53. Voisin 1987: 518–41 saw the cause in Nero's taste for things Egyptian, and specifically Alexandrian.
8 Hiesinger 1975; L'Orange 1947.
9 Lehmann 1945.
10 Recent interpretations of the Pantheon have emphasised this celestial symbolism (Stierlin 1984: 73–111, 1986: 123–34), and encourage us to take it back to Agrippa's original structure (Thomas 1997, 2004).
11 Potamianos 2000.
12 Heilbron 1999; cf. Hannah 2007a.
13 Heilbron 1999: 288–89.
14 It is a pity that the desire to harness the sunlight for another, more mundane purpose, that of 'daylight saving' since the mid-1970s, has caused the curators of the Shrine to instal mirrors to ensure the continuation of the display at the intended time.
15 For further thoughts on this mode of thinking and its contrast with our modern understanding of the mechanics of the cosmos, see Hannah 2006a.

REFERENCES

Ancient authors

Note: In all but one case, the translations of Classical authors in this book are my own, based on the original texts. This Bibliography presents editions of these texts which are readily accessible. The abbreviation *LCL* indicates a volume in the *Loeb Classical Library*, which provides the Greek or Latin text with English translations on facing pages. By convention, authors' names in square brackets indicate disputed or spurious authorship.

Aelian, *Aelian, Historical Miscellany, vol. 4*, trans. N. G. Wilson, Cambridge, Mass.: Harvard University Press (*LCL*), 1997.

Aineias the Tactician, *How to Survive under Siege*, trans. D. Whitehead, Oxford: Clarendon Press; New York: Oxford University Press, 1990.

Aiskhines, *Aeschines, Speeches*, trans. C. D. Adams, Cambridge, Mass.: Harvard University Press (*LCL*), 1919.

Aiskhylos, *Aeschylus*, vols 1 and 2, trans. A. H. Sommerstein, Cambridge, Mass.: Harvard University Press (*LCL*), forthcoming 2009.

Alkiphron, *Alciphron, Aelian, and Philostratus, The Letters*, trans. A. R. Benner and F. H. Fobes, London: Heinemann, (*LCL*), 1949.

Aratos, *Phainomena*, in *Callimachus, Hymns and Epigrams; Lycophron; Aratus*, trans. A. W. Mair and G. R. Mair, Cambridge, Mass.: Harvard University Press (*LCL*), 1921.

Aristophanes, in *Aristophanes*, trans. J. Henderson, 5 vols, Cambridge, Mass.: Harvard University Press (*LCL*), 1998–2008.

[Aristotle], *The Athenian Constitution*, in *The Athenian Constitution; The Eudemian Ethics; On Virtues and Vices*, trans. H. Rackham, Cambridge, Mass.: Harvard University Press, 1952.

Aristotle, *On the Heavens*, in *Aristotle*, vol. 6, trans. W. K. C. Guthrie, Cambridge, Mass.: Harvard University Press (*LCL*), 1971.

Aristotle, *Physics*, in *Aristotle, vols 4–5, The Physics*, trans. P. H. Wicksteed and F. M. Cornford, Cambridge, Mass.: Harvard University Press (*LCL*), 1934.

Aristotle, *Poetics*, in *Aristotle, vol. 23, Poetics, On the Sublime, On Style*, trans. S. Halliwell, W. H. Fyfe, D. C. Innes, revised by D. A. Russell, Cambridge, Mass.: Harvard University Press (*LCL*), 1932.

[Aristotle], *Problems*, in *Aristotle, vol. 15, Problems, Books 1–21*, trans. W. S. Hett, Cambridge, Mass.: Harvard University Press (*LCL*), 1961.

Artemidoros, in *Artemidoros, The Interpretation of Dreams = Oneirocritica*, trans. R. J. White, Park Ridge, N.J.: Noyes Press, 1975.
Asconius, *For Milo*, in *Commentaries on Five Speeches of Cicero, Asconius Pedianus, Quintas*, trans. S. Squires, Bristol: Bristol Classical Press; Wauconda, Ill.: Bolchazy–Carducci, 1990.
Athenaios, in *Athenaeus, The Learned Banqueters*, 7 vols, trans. S. D. Olson, C. B. Gulick, Cambridge, Mass.: Harvard University Press (*LCL*), 1969–1987.
Augustine, *Confessions*, in *Augustine, Confessions*, 2 vols, trans. W. Watts, Cambridge, Mass.: Harvard University Press (*LCL*), 1912.
Aulus Gellius, in *Aulus Gellius, Attic Nights*, 3 vols, trans. J. C. Rolfe, Cambridge, Mass.: Harvard University Press (*LCL*), 1927.
Caesar, The Gallic War, trans. H. J. Edwards, Cambridge, Mass.: Harvard University Press (*LCL*), 1917.
Cassiodorus, *Letters*, in *The Variae of Magnus Aurelius Cassiodorus Senator*, trans. S. J. B. Barnish, Liverpool: Liverpool University Press, 1992.
Cassius Dio, in *Dio Cassius*, 9 vols, trans. E. Cary, H. B. Foster, Cambridge, Mass.: Harvard University Press (*LCL*), 1914–1927.
Censorinus, *On the Birthday*, in *Censorinus, The Birthday Book*, trans. H. N. Parker, Chicago, Ill.: University of Chicago Press, 2007.
Cetius Faventinus, in *Cetius Faventinus, Abrégé d'architecture privée*, trans. M.-T. Cam, Paris: Les Belles Lettres, 2001.
Cicero, *Brutus*, in *Cicero, vol. 5, Rhetorical Treatises*, trans. G. L. Hendrickson and H. M. Hubbell, Cambridge, Mass.: Harvard University Press (*LCL*), 1939.
Cicero, *For Quinctius*, in *Cicero, vol. 6, Orations*, trans. J. H. Freese, Cambridge, Mass.: Harvard University Press (*LCL*), 1930.
Cicero, *For Rabirius*, in *Cicero, vol. 9, Orations*, trans. H. G. Hodge, Cambridge, Mass.: Harvard University Press (*LCL*), 1927.
Cicero, *Letters to Atticus*, in *Cicero, vols. 22–24, 29, Letters to Atticus*, trans. D. R. Shackleton-Bailey, Cambridge, Mass.: Harvard University Press (*LCL*), 1999.
Cicero, *Republic*, in *Cicero, vol. 16, Philosophical Treatises*, trans. C. W. Keyes, Cambridge, Mass.: Harvard University Press (*LCL*), 1928.
Cicero, *The Nature of the Gods*, in *Cicero, vol. 19, Philosophical Treatises*, trans. H. Rackham, Cambridge, Mass.: Harvard University Press (*LCL*), 1933.
Cicero, *Tusculan Disputations*, in *Cicero, vol. 18, Philosophical Treatises*, trans. J. E. King, Cambridge, Mass.: Harvard University Press (*LCL*), 1927.
Columella, *On Agriculture*, in *Columella, On Agriculture*, 3 vols, trans. H. B. Ash, E. S. Forster, E. H. Heffner, Cambridge, Mass.: Harvard University Press (*LCL*), 1941–1955.
Cornelius Nepos, in *Cornelius Nepos, On Great Generals, On Historians*, trans. J. C. Rolfe, Cambridge, Mass.: Harvard University Press (*LCL*), 1929.
Demosthenes, in *Demosthenes*, 7 vols, trans. J. H. Vince, C. A. Vince, A. T. Murray, N. W. de Witt, Cambridge, Mass.: Harvard University Press (*LCL*), 1926–1949.
Diodoros Siculus, in *Diodorus Siculus, Library of History*, 12 vols, trans. C. H. Oldfather, C. L. Sherman, C. B. Welles, R. M. Geer, F. R. Walton, Cambridge, Mass.: Harvard University Press (*LCL*), 1933–1967.
Diogenes Laertius, in *Diogenes Laertius, Lives of Eminent Philosophers*, 2 vols, trans. R. D. Hicks, Cambridge, Mass.: Harvard University Press (*LCL*), 1925.

Dioscorides Pedanius, in M. Wellmann (ed.), *Pedanii Dioscuridis Anazarbei de materia medica libri quinque*, 3 vols, Berlin: Weidmann, 1906–1914 (repr. 1958).

Euripides, *Chrysippos*, in *Euripides, vol. 8, Fragments*, trans. C. Collard and M. Cropp, Cambridge, Mass.: Harvard University Press (*LCL*), forthcoming.

Euripides, *Orestes*, in *Euripides, vol. 5*, trans. D. Kovacs, Cambridge, Mass.: Harvard University Press (*LCL*), 2002.

[Euripides], *Rhesos*, in *Euripides, vol. 6*, trans. D. Kovacs, Cambridge, Mass.: Harvard University Press (*LCL*), 2003.

Geminos, *Introduction to Astronomy* in *Geminos's Introduction to the Phenomena: A Translation and Study of a Hellenistic Survey of Astronomy*, trans. and commentary J. Evans and J. L. Berggren, Princeton, N.J.: Princeton University Press, 2006.

Harpokration, in *Harpocration, Lexeis of the Ten Orators*, ed J. J. Keaney, Amsterdam: Adolf M. Hakkert, 1991.

Herodotos, in *Herodotus, The Persian Wars*, 4 vols, trans. A. D. Godley, Cambridge, Mass.: Harvard University Press (*LCL*), 1920–1925.

Hesiod, *Works and Days*, in *Hesiod, vol. 1*, trans. G. W. Most, Cambridge, Mass.: Harvard University Press (*LCL*), 2007.

Hipparkhos, *Commentary on the Phaenomena of Aratos and Eudoxos*, in C. Manitius (ed.), *Hipparchi in Arati et Eudoxi phaenomena commentariorum libri iii*. Leipzig: Teubner, 1894.

Hipparkhos, *Geographical Fragments*, in D. R. Dicks (ed.), *The Geographical Fragments of Hipparchus*. London: Athlone Press, 1960.

[Hippokrates], *On Airs, Waters, and Places*, in *Hippocrates, vol. 1*, trans. W. H. S. Jones, Cambridge, Mass.: Harvard University Press (*LCL*), 1923.

Hippokrates, *On Flesh*, in *Hippocrates, vol. 8*, trans. P. Potter, Cambridge, Mass.: Harvard University Press (*LCL*), 1995.

[Hippokrates], *On Regimen*, in *Hippocrates, vol. 4*, trans. W. H. S. Jones, Cambridge, Mass.: Harvard University Press (*LCL*), 1931.

Homer, *Odyssey*, in *Homer, vols 3–4, Odyssey*, trans. A. T. Murray, revised by G. E. Dimock, Cambridge, Mass.: Harvard University Press (*LCL*), 1998.

Horace *Odes*, in *Horace, vol. 1*, trans. N. Rudd, Cambridge, Mass.: Harvard University Press (*LCL*), 2004.

Horace *Satires*, in *Horace, vol. 2*, trans. H. R. Fairclough, Cambridge, Mass.: Harvard University Press (*LCL*), 1926.

Juvenal, in *Juvenal and Persius*, trans. S. M. Braund, Cambridge, Mass.: Harvard University Press (*LCL*), 2004.

Kallimachos, in *Callimachus, vol. 1*, trans. C. A. Trypanis, T. Gelzer, C. H. Whitman, Cambridge, Mass.: Harvard University Press (*LCL*), 1973.

Kleomedes, *The Heavens* in *Cleomedes' Lectures on Astronomy: A Translation of the Heavens with an Introduction and Commentary*, eds A. C. Bowen and R. B. Todd, Berkeley: University of California Press, 2004.

Livy, Book 31 in *Livy, History of Rome, vol. 9*, trans. E. T. Sage, Cambridge, Mass.: Harvard University Press (*LCL*), 1935.

Lucan, in *Lucan, The Civil War*, trans. J. D. Duff, Cambridge, Mass.: Harvard University Press (*LCL*), 1928.

Lucian, in *Lucian*, 8 vols, trans. A. M. Harmon, K. Kilburn and M. D. MacLeod, Cambridge, Mass.: Harvard University Press (*LCL*), 1913–1967.

Macrobius, *Saturnalia*, in *Macrobius, The Saturnalia*, trans. P. V. Davies, New York, Columbia University Press, 1969.

Martial, *Epigrams* in *Martial*, 3 vols, trans. D. R. Shackleton Bailey, Cambridge, Mass.: Harvard University Press (*LCL*), 1993.

Pappos, in *Commentaires de Pappus et de Théon d'Alexandrie sur l'Almageste*, ed. A. Rome, 3 vols, Rome: Biblioteca apostolica vaticana, 1931–1943.

Petronius, *Satyricon* in *Petronius, Satyricon, Seneca, Apocolocyntosis*, trans. M. Heseltine, W. H. D. Rouse and E. H. Warmington, Cambridge, Mass.: Harvard University Press (*LCL*), 1913.

Pindar, *Olympian Odes*, in *Pindar, vol. 1*, trans. W. H. Race, Cambridge, Mass.: Harvard University Press (*LCL*), 1997.

Plato, *Kritias*, in *Plato, vol. 9*, trans. R. G. Bury, Cambridge, Mass.: Harvard University Press (*LCL*), 1929.

Plato, *Laws*, in *Plato, vols 10–11*, trans. R. G. Bury, Cambridge, Mass.: Harvard University Press (*LCL*), 1926.

Plato, *Protagoras*, in *Plato, vol. 2*, trans. W. R. M. Lamb, Cambridge, Mass.: Harvard University Press (*LCL*), 1924.

Plato, *Theaetetus*, in *Plato, vol. 7*, trans. H. N. Fowler, Cambridge, Mass.: Harvard University Press (*LCL*), 1921.

Plato, *Timaeus*, in *Plato, vol. 9*, trans. R. G. Bury, Cambridge, Mass.: Harvard University Press (*LCL*), 1929.

Plautus, *Amphitruo*, in *Plautus, vol. 1*, trans. P. Nixon, Cambridge, Mass.: Harvard University Press (*LCL*), 1916.

Plautus, *Pseudolus*, in *Plautus, vol. 4*, trans. P. Nixon, Cambridge, Mass.: Harvard University Press (*LCL*), 1932.

Pliny the Elder, *Natural History*, trans. H. Rackham and D. E. Eichholz, 10 vols, Cambridge, Mass.: Harvard University Press (*LCL*), 1938–1962.

Pliny the Younger, *Letters*, trans. B. Radice, 2 vols, Cambridge, Mass.: Harvard University Press (*LCL*), 1969.

Plutarch, *Alexander*, in *Plutarch, Parallel Lives, vol. 7*, trans. B. Perrin, Cambridge, Mass.: Harvard University Press (*LCL*), 1919.

Plutarch, *Camillus*, in *Plutarch, Parallel Lives, vol. 2*, trans. B. Perrin, Cambridge, Mass.: Harvard University Press (*LCL*), 1914.

Plutarch, *Solon*, in *Plutarch, Parallel Lives, vol. 1*, trans. B. Perrin, Cambridge, Mass.: Harvard University Press (*LCL*), 1914.

Polybios, *Histories*, in *Polybius, Histories*, trans. W. R. Paton, 6 vols, Cambridge, Mass.: Harvard University Press (*LCL*), 1922–1927.

Proklos, in Friedlein, G. 1873. *Procli Diadochi in primum Euclidis elementorum librum commentarii*. Leipzig: Teubner.

Ptolemy, *Almagest*, trans. G. J. Toomer, Princeton, N.J.: Princeton University Press, 1998.

Ptolemy, *Phaseis*, text and translation in Lehoux 2007: 261–309.

Ptolemy, *Tetrabiblos*, trans. F. E. Robbins, Cambridge, Mass.: Harvard University Press (*LCL*), 1940.

Rule of the Master in *The Rule of the Master / Regula magistri*, trans. L. Eberle and C. Philippi. Kalamazoo: Cistercian Publications, 1977.

Seneca, *Apocolocyntosis*, in *Petronius, Satyricon, Seneca, Apocolocyntosis*, trans. M. Heseltine,

W. H. D Rouse and E. H. Warmington, Cambridge, Mass.: Harvard University Press (*LCL*), 1913.

Sextus Empiricus, *Against the Professors*, trans. R. G. Bury, Cambridge, Mass.: Harvard University Press (*LCL*), 1949.

Simplicius, *On Aristotle's On the Heavens*, in *Simplicius, On Aristotle on the Heavens 1. 5–9*, trans. R. J. Hankinson, London: Duckworth, 2003; and in *Simplicius, On Aristotle on the Heavens 2. 10–14*, trans. I. Mueller, London: Duckworth, 2005.

Soranos, in *Soranos d'Ephèse, Maladies des femmes*, eds P. Burguière, Danielle Gourevitch and Yves Malinas, Paris: Les Belles Lettres, 1988–1994.

Strabo, *Geography*, trans. H. L. Jones, 8 vols, Cambridge, Mass.: Harvard University Press (*LCL*), 1917–1932.

Suetonius, in *Suetonius, The Lives of the Caesars*, trans. K. R. Bradley and J. C. Rolfe, Cambridge, Mass.: Harvard University Press (*LCL*), 1914.

Tacitus, *Dialogue on Orators*, in *Tacitus*, vol. 1, trans. M. Hutton *et al.*, Cambridge, Mass.: Harvard University Press (*LCL*), 1970.

[Theophrastos] *On Weather-Signs*, in *Theophrastus, vol. 2*, trans. A. F. Hort, Cambridge, Mass.: Harvard University Press (*LCL*), 1916.

Thucydides, in *Thucydides, History of the Peloponnesian War*, trans. C. F. Smith, 2 vols, Cambridge, Mass.: Harvard University Press (*LCL*), 1919–1920.

Varro, *On Farming*, in *Varro, Cato, On Agriculture*, trans. W. D. Hooper and H. B. Ash, Cambridge, Mass.: Harvard University Press (*LCL*), 1934.

Varro, *On the Latin Language*, trans. R. G. Kent, 2 vols, Cambridge, Mass.: Harvard University Press (*LCL*), 1938.

Vitruvius, *On Architecture*, trans. F. Granger, 2 vols, Cambridge, Mass.: Harvard University Press (*LCL*), 1931, 1934.

Xenophon, *Cyropaedia*, in *Xenophon, vol. 5*, trans. W. Miller, Cambridge, Mass.: Harvard University Press (*LCL*), 1914.

Modern authors

Abry, J.-H. (ed.) 1993. *Les tablettes astrologiques de Grand (Vosges) et l'astrologie en Gaule romaine. Actes de la Table-Ronde du 18 mars 1992*, Paris: de Boccard.

Akurgal, E. 1993. *Ancient Civilizations and Ruins of Turkey*, 8th ed., Istanbul: NET Turistik Yayinlar.

Allen, D. 1996. 'A Schedule of Boundaries: An Exploration, Launched from the Water-clock, of Athenian Time', *Greece and Rome* 43: 157–68.

Ardaillon, E. 1900. 'Horologium', in C. Daremberg and E. Saglio (eds), *Dictionnaire des antiquités grecques et romaines d'après les textes et les monuments*, Paris: Hachette, vol. 3, 256–64.

Armstrong, J. E. and J. McK. Camp 1977. 'Notes on a Water Clock in the Athenian Agora', *Hesperia* 46: 147–61.

Arnaldi, M. and K. Schaldach 1997. 'A Roman Cylinder Dial: Witness to a Forgotten Tradition', *Journal for the History of Astronomy* 28: 107–17.

Arnaud, P. 1984. 'L'image du globe dans le monde romain: science, iconographie, symbolique', *Mélanges de l'École française de Rome: Antiquité* 96: 53–116.

Arnaud, P. 2007. 'Diocletian's Prices Edict: The Prices of Seaborne Transport and the Average Duration of Maritime Travel', *Journal of Roman Archaeology* 20: 321–36.

Ashmore, T. 2007a. 'Astrolabes, Part 1: Introduction', *Bulletin of the British Sundial Society* 19: 91–95.

Ashmore, T. 2007b. 'Astrolabes, Part 2: European Astrolabes', *Bulletin of the British Sundial Society* 19: 107–11.

Ashmore, T. 2007c. 'Astrolabes, Part 3: Arabic Astrolabes', *Bulletin of the British Sundial Society* 19: 156–60.

Ashmore, T. 2008. 'Astrolabes, Part 4: Universal Astrolabes', *Bulletin of the British Sundial Society* 20: 31–37.

Barton T. 1994. *Ancient Astrology*, London: Routledge.

Beard, M., J. North and S. Price 1998. *Religions of Rome*, Cambridge: Cambridge University Press.

Beck, R. 2006. *The Religion of the Mithras Cult in the Roman Empire: Mysteries of the Unconquered Sun*, Oxford: Oxford University Press.

Beck, R. 2007. *A Brief History of Ancient Astrology*, Oxford: Blackwell.

Belmonte, J. A. 2003. 'The Ramesside star clocks and the ancient Egyptian constellations', in M. Blomberg, P. E. Blomberg and G. Henriksson (eds), *Calendars, Symbols, and Orientations: Legacies of Astronomy in Culture*, Uppsala, Uppsala Astronomical Observatory, 57–65.

Bendorff, O., E. Weiss and A. Rehm 1903. 'Zur Salzburger Bronzescheibe mit Sternbildern', *Jahreshefte des Österreichischen Archäologischen Institutes in Wien* 6: 32–49.

Bennett, C., 2004. 'The Early Augustan Calendars in Rome and Egypt: Addenda et Corrigenda', *Zeitschrift für Papyrologie und Epigraphik* 147: 165–68.

Bennett, C., forthcoming 2009, 'Alexandria and the Ptolemaic Macedonian Calendar', *Ancient Society* 39.

Best, E. 1922. *The Astronomical Knowledge of the Maori, Genuine and Empirical*, Wellington: Government Printer.

Betz, H. D. (ed.) 1986. *The Greek Magical Papyri in Translation including the Demotic Spells*, Chicago: University of Chicago Press.

Bickerman, E. J. 1980. *Chronology of the Ancient World*, Revised ed., London: Thames and Hudson.

Bilfinger, G. 1886. *Die Zeitmesser der antiken Völker*, Stuttgart: K. Hofbuchdruckerei Carl Liebich.

Blackburn, B. and L. Holford-Strevens 2003. *The Oxford Companion to the Year: An Exploration of Calendar Customs and Time-reckoning*, Oxford: Oxford University Press.

Blass, F. 1887. *Eudoxi Ars astronomica qualis in charta Aegyptiaca superest denuo edita*, Kiel: Schmidt & Klaunig.

Bobrova, L. and A. Militarev 1993. 'From Mesopotamia to Greece: On the Origin of Semitic and Greek Star Names', in H. D. Galter (ed.), *Die Rolle der Astronomie in den Kulturen Mesopotamiens: Beiträge zum 3. Grazer Morgenländischen Symposion (23.–27. September 1991)*, Graz: rm-Druck & Verlagsgesellschaft, 307–30.

Borchardt, L. 1920. *Die altägyptische Zeitmessung. Die Geschichte der Zeitmessung und der Uhren*, Band 1, Lieferung B, Berlin: Vereinigung Wissenschaftlicher Verleger Walter de Gruyter.

Boutsikas, E. 2007. 'Astronomy and Ancient Greek Cult: An Application of Archaeoastronomy to Greek ReligiousArchitecture, Cosmologies and Landscapes', Unpublished PhD thesis, Leicester: University of Leicester.

Bowen, A. C. 2006. 'Geminus and the Length of the Month: The Authenticity of Intro. Ast. 8. 43–45', *Journal for the History of Astronomy* 37: 193–202.

Bowen, A. C. and B. R. Goldstein 1988. 'Meton of Athens and astronomy in the late fifth century B.C.', in E. Leichty, M. de J. Ellis, and P. Gerardi (eds), *A Scientific Humanist: Studies in Memory of Abraham Sachs*, Philadelphia, Penn.: The University Museum, 39–81.

Bowen, A. C. and B. R. Goldstein 1991. 'Hipparchus' Treatment of Early Greek Astronomy: The Case of Eudoxus and the Length of Daytime', *Proceedings of the American Philosophical Society* 135: 233–54.

Bowen, A. C. and R. B. Todd (eds), 2004. *Cleomedes' Lectures on Astronomy: A Translation of The Heavens with an Introduction and Commentary*, Berkeley: University of California Press.

Braudel, F. 1979. *Civilization and Capitalism, 15th–18th Century, Volume 1, The Structures of Everyday Life: The Limits of the Possible*, New York: Harper & Row.

Bremner, R. W. 1993. 'The Shadow Length Table in Mul.Apin', in H. D. Galter (ed.), *Die Rolle der Astronomie in den Kulturen Mesopotamiens: Beiträge zum 3. Grazer Morgenländischen Symposion (23.–27. September 1991)*, Graz: rm-Druck & Verlagsgesellschaft, 367–82.

Bromley, A. G. 1986. 'Notes on the Antikythera Mechanism', *Centaurus* 29: 5–27.

Bromley, A. G. 1990. 'Observations of the Antikythera Mechanism', *Antiquarian Horology* 18: 641–52.

Buchner, E. 1976. 'Solarium Augusti und Ara Pacis', *Römische Mitteilungen* 83: 319–65.

Buchner, E. 1980. 'Horologium Solarium Augusti: Vorbericht über die Ausgrabungen 1979/80', *Römische Mitteilungen* 87: 355–73.

Buchner, E. 1982. *Die Sonnenuhr des Augustus: Nachdruck aus RM 1976 und 1980 und Nachtrag über die Ausgrabung 1980/1981*, Mainz am Rhein: von Zabern.

Buchner, E. 1993–94. 'Neues zur Sonnenuhr des Augustus', *Nürnberger Blätter zur Archäologie* 10: 77–84.

Buxton, B., and R. Hannah 2005. 'OGIS 458, the Augustan Calendar, and the Succession', in C. Deroux (ed.), *Studies in Latin Literature and Roman History*, vol. XII, Collection Latomus 287, Brussels: Latomus, 290–306.

Cabanes, P. 2003. 'Recherches sur le calendrier corinthien en Épire et dans les régions voisines', *Revue des Études Anciennes* 105: 83–102.

Cam, M.-T. 2001. *Cetius Faventinus, Abrégé d'architecture privée*, Paris: Les Belles Lettres.

Camp, J. McK. 1992. *The Athenian Agora: Excavations in the Heart of Classical Athens*, London: Thames and Hudson.

Casson, L. 1994. *Travel in the Ancient World*, Baltimore, Md.: Johns Hopkins University Press.

Cavanaugh, M. B., 1996. *Eleusis and Athens: Documents in Finance, Religion, and Politics in the Fifth Century B.C.*, Atlanta, Ga.: Scholars Press.

Cotterell, B., F. P. Dickson and J. Kamminga 1986. 'Ancient Egyptian Water-clocks: A Reappraisal', *Journal of Archaeological Science* 13: 31–50.

Crawford, M. H. (ed.) 1996. *Roman Statutes*, London: Institute of Classical Studies, School of Advanced Study, University of London.

Davidson, N. 1985. *Astronomy and the Imagination: A New Approach to Man's Experience of the Stars*, London: Routledge and Kegan Paul.

Davies, J. K., 2003. 'Greek Archives: From Record to Monument', in M. Brosius (ed.), *Ancient Archives and Archival Traditions: Concepts of Record-keeping in the Ancient World*, Oxford, New York: Oxford University Press, 323–43.

De Fine Licht, K. 1968. *The Rotunda in Rome: A Study of Hadrian's Pantheon*, Jutland Archaeological Society publications, 8. Copenhagen: Gyldendal.

De Vos, A. and M. De Vos 1982. *Pompei, Ercolano, Stabia*, Roma: G. Laterza.

Delambre, J. B. J. 1817. *Histoire de l'astronomie ancienne*, Tome 2, Paris: V. Courcier.

Dicks, D. R. 1966. 'Solstices, Equinoxes, and the Presocratics', *Journal of Hellenic Studies* 86: 26–40.

Dicks, D. R. 1970. *Early Greek Astronomy to Aristotle*, London: Thames and Hudson.

Diels, H. 1924/1965. *Antike Technik: sieben Vorträge*. Osnabrück: Zeller, 1965, reprint of 1924 edition.

Diels, H. and A. Rehm 1904. 'Parapegmenfragmente aus Milet', *Sitzungsberichte der königlich preussischen Akademie der Wissenschaften, philosophisch-historischen Classe* 23, 92–111.

Diggle, J. 1998. *Tragicorum Graecorum fragmenta selecta*, Oxford: Clarendon Press.

Dilke, O. A. W. 1985. *Greek and Roman Maps*, London: Thames and Hudson.

Dilke, O. A. W. 1987. *Mathematics and Measurement*, London: British Museum Publications.

Dillon, J. and T. Gergel 2003. *The Greek Sophists*, London, New York: Penguin.

Dindorf, G. 1962. *Tragoediae superstites et deperditarum fragmenta*, Hildesheim: Olm.

Dinsmoor, W. B. 1931/1966. *The archons of Athens in the Hellenistic age*, Amsterdam: A. M. Hakkert.

Dohrn-van Rossum, G. 1996. *History of the Hour: Clocks and Modern Temporal Orders*, trans. T. Dunlap, Chicago, Ill.: University of Chicago Press.

Dohrn-van Rossum, G. 2003. 'Water Clocks', in H. Cancik and H. Schneider (eds), *Brill's New Pauly: Encyclopaedia of the Ancient World*, Leiden: Brill, 462–64.

Dover, K. J. (ed.) 1968. *Aristophanes: Clouds*, Oxford: Clarendon Press.

Dow, S. 1968. 'Six Athenian Sacrificial Calendars', *Bulletin de Correspondance Hellénique* 92: 170–86.

Drachmann, A. G. 1954. 'The Plane Astrolabe and Anaphoric Clock', *Centaurus* 3: 183–89.

Drecker, J. 1925. *Die Theorie der Sonnenuhren. Die Geschichte der Zeitmessung und der Uhren*, Band 1, Lieferung E, Berlin and Leipzig: Vereinigung Wissenschaftlicher Verleger Walter de Gruyter.

Dumser, E. A. 2002. 'Pantheum', in L. Haselberger *et al.*, *Mapping Augustan Rome*, Portsmouth, R.I.: Journal of Roman Archaeology, Supplementary Series no. 50, 188–89.

Dunbar, N. 1995. *Aristophanes: Birds*, Oxford: Clarendon Press.

Dunn, F. M., 1999. 'Tampering with the Calendar', *Zeitschrift für Papyrologie und Epigraphik* 123: 213–31.

Edwards, D. 1984. 'Ptolemy's Peri analemmatos: An Annotated Transcription of Moerbeke's Latin Translation and of the Surviving Greek Fragments with an English Version and Commentary', Unpublished PhD thesis, Providence, R.I.: Brown University.

Eitrem, S. 1991. 'Dreams and Divination in Magical Ritual', in C. A. Faraone and D. Obbink (eds), *Magika Hiera: Ancient Greek Magic and Religion*, New York: Oxford University Press.

Eliot, C. W. J. 1976. 'Oropos (Skala Oropou), Greece', in R. Stillwell (ed.), *The Princeton Encyclopedia of Classical Sites*, Princeton, N.J.: Princeton University Press.

Evans, J. 1999. 'The Material Culture of Greek Astronomy', *Journal for the History of Astronomy* 30: 237–307.

Evans, J. and J. L. Berggren (eds) 2006. *Geminos's Introduction to the Phenomena: A Translation and Study of a Hellenistic Survey of Astronomy*, Princeton, N.J.: Princeton University Press.

Evans, J. and M. Marée 2008. 'A Miniature Ivory Sundial with Equinox Indicator from Ptolemaic Tanis, Egypt', *Journal for the History of Astronomy* 39: 1–17.

Evans, J. 1998. *The History and Practice of Ancient Astronomy*, Oxford: Oxford University Press.

Evans, N. A. 2002. 'Sanctuaries, Sacrifices, and the Eleusinian Mysteries', *Numen* 49: 227–54.

Feeney, D. 2007. *Caesar's Calendar: Ancient Time and the Beginnings of History*, Berkeley: University of California Press.

Fornara, C. W. 1983. *Archaic Times to the End of the Peloponnesian War*, Cambridge: Cambridge University Press.

Frederiksen, M. 1984. *Campania*. London: British School at Rome.

Freeth, T., A. Jones, J. M. Steele and Y. Bitsakis 2008. 'Calendars with Olympiad Display and Eclipse Prediction on the Antikythera Mechanism', *Nature* 454 (31 July 2008): 614–17.

Freeth, T., Y. Bitsakis, X. Moussas, J. H. Seiradakis, A. Tselikas, E. Magkou, M. Zafeiropoulou, R. Hadland, D. Bate, A. Ramsey, M. Allen, A. Crawley, P. Hockley, T. Malzbender, D. Gelb, W. Ambrisco, and M. G. Edmunds 2006. 'Decoding the Ancient Greek Astronomical Calculator known as the Antikythera Mechanism', *Nature* 444 (30 November 2006): 587–91.

Freidel, D., L. Schele and J. Parker 1993. *Maya Cosmos: Three Thousand Years on the Shaman's Path*, New York: W. Morrow.

Friberg, J., H. Hunger and F. N. H. Al–Rawi 1990. 'Seed and Reeds – A Metro-mathematical Topic Text from Late Babylonian Uruk', *Baghdader Mitteilungen* 21: 483–557.

Gatty, Mrs Alfred 1890. *The Book of Sun–dials*, 3rd ed., London: George Bell and Sons.

Gibbs, S. L. 1976. *Greek and Roman Sundials*, New Haven, Conn.: Yale University Press.

Gibbs, S. L. 1979. 'The First Scientific Instruments', in K. Brecher and M. Feirtag (eds), *Astronomy of the Ancients*, Cambridge, Mass.: MIT Press, 39–59.

Ginzel, F. K., 1911, *Handbuch der mathematischen und technischen Chronologie*, vol. 2, Leipzig: J.C. Hinrichs'sche Buchhandlung.

Grainger, F. 1933. 'The Greek Origin of the Pantheon', *Journal of the Royal Institute of British Architects* 40: 57–61.

Greenidge, A. H. J. 1901/1971. *The Legal Procedure of Cicero's Time*, reprint of the 1901 ed., New York: A. M. Kelley.

Grenfell, B. P. and A. S. Hunt (eds) 1906. *The Hibeh Papyri*, vol. 1, London: Egypt Exploration Fund.

Griffith, M. 1983. *Aeschylus, Prometheus Bound*. Cambridge: Cambridge University Press.

Griffiths, J. G. (ed.) 1970. *Plutarch: De Iside et Osiride*, Cardiff: University of Wales Press.

Gros, P. 1996. *L'architecture romaine: du début du IIIe siècle av. J.-C. à la fin du Haut-Empire*, vol. 1, Paris: Picard.

Gundel, H. G. 1992. *Zodiakos: Tierkreisbilder im Altertum: Kosmische Bezüge und Jenseitsvorstellungen im antiken Alltagsleben*, Mainz am Rhein: P. von Zabern.
Habicht, C. 1999. *Athens from Alexander to Antony*, trans. D. L. Schneider, Cambridge, Mass.: Harvard University Press.
Hall, J. C. R. 2006. *Performing Cicero's Pro Archia* (video recording). New Zealand.
Hannah, R. 1989. '... *praevolante nescio qua ingenti humana specie* ... A Reassessment of the Winged Genius on the Base of the Antonine Column', *Papers of the British School at Rome* 57: 90–105.
Hannah, R. 1993a. 'Alcumena's Long Night: Plautus, *Amphitruo* 273–76', *Latomus* 52: 65–74.
Hannah, R. 1993b. 'The Stars of Iopas and Palinurus', *American Journal of Philology* 114: 123–35.
Hannah, R. 1997a. 'The Temple of Mars Ultor and 12 May', *Römische Mitteilungen* 104: 374–86.
Hannah, R. 1997b. 'Is it a Bird? Is it a Star? Ovid's Kite and the First Swallow of Spring', *Latomus* 56: 327–42.
Hannah, R. 1997c. 'Odysseus's Navigation', *Prudentia* 29: 15–33.
Hannah, R. 1998. 'Games for Mars and the Temples of Mars Ultor', *Klio* 80: 422–33.
Hannah, R. 2001a. 'From Orality to Literacy? The Case of the Parapegma', in J. Watson (ed.), *Speaking Volumes: Orality and Literacy in the Greek and Roman World*, Leiden: Brill, 139–59.
Hannah, R. 2001b. 'The Moon, the Sun and the Stars: Counting the Days and the Years', in S. McCready (ed.), *The Discovery of Time*. London: MQ Publications, 56–99.
Hannah, R. 2002. 'Euctemon's Parapegma', in C. J. Tuplin and T. E. Rihll (eds), *Science and Mathematics in Ancient Greek Culture*, Oxford: Oxford University Press, 112–32.
Hannah, R. 2005. *Greek and Roman Calendars: Constructions of Time in the Classical World*, London: Duckworth.
Hannah, R. 2006a. 'The Unconquered Sun', in Tom Barnes *et al.*, *The Elegant Universe of Albert Einstein*, Wellington: Awa Press, 119–43.
Hannah, R. 2006b. 'Parapegmata, Hellenistic Calendars and the Antikythera Mechanism', paper delivered at *Decoding the Antikythera Mechanism: Science and Technology in Ancient Greece*, Athens.
Hannah, R. 2007a. 'The Meridiana of Ulugh Beg in Hagia Sophia, Constantinople', *Nuncius* 22: 1–14.
Hannah, R. 2007b. 'Between Science and Literature: Star Calendars in the Greek and Roman Worlds', paper delivered at the *Leeds International Classics Seminar 2007*, Leeds.
Hannah, R. 2008. 'Timekeeping', in J. P. Oleson (ed.), *Handbook of Engineering and Technology in the Classical World*, Oxford: Oxford University Press, 740–58.
Hardy, E. G. (ed.) 1912. *Roman Laws and Charters*, Oxford: Clarendon Press.
Harris, W. V. 1989. *Ancient Literacy*, Cambridge, Mass.: Harvard University Press.
Harvey, B. K. 2004. *Roman Lives: Ancient Roman Life as Illustrated by Latin Inscriptions*, Newburyport, Mass.: Focus Pub./R. Pullins Co.
Heath, T. L. 1956. *The Thirteen Books of Euclid's Elements*, reprint of 2nd ed. New York: Dover Publications.
Heilbron, J. L. 1999. *The Sun in the Church: Cathedrals as Solar Observatories*, Cambridge, Mass. and London: Harvard University Press.

Henriksson, G. 2007. 'Orientation of 140 Swedish Passage graves: A Megalithic Calendar', in M. P. Zedda and J. A. Belmonte (eds), *Lights and Shadows in Cultural Astronomy. Proceedings of the SEAC 2005, Isili, Sardinia 28 June to 3 July*. Sardinia: Associazione Archeofila Sarda, 289–297.

Heslin, P. 2007. 'Augustus, Domitian and the So-called Horologium Augusti', *Journal of Roman Studies* 97: 1–20.

Hetland, L. M. 2007. 'Dating the Pantheon', *Journal of Roman Archaeology* 20: 95–112.

Heyerdahl, T. and E. N. Ferdon (eds) 1961. *Reports of the Norwegian Archaeological Expedition to Easter Island and the East Pacific, i. Archaeology of Easter Island*, London: Allen and Unwin.

Hicks, E. L. 1887. 'Iasos', *Journal of Hellenic Studies* 8: 83–118.

Hiesinger, U. W. 1975. 'The Portraits of Nero', *American Journal of Archaeology* 79: 113–24.

Hornblower, S. 1991. *A Commentary on Thucydides, Volume I, Books I–III*. Oxford: Clarendon Press.

Hoskin, M. and O. Gingerich 1997. 'Astronomy's Widening Horizons', in M. Hoskin (ed.), *The Cambridge Illustrated History of Astronomy*. Cambridge: Cambridge University Press, 344–65.

Houston, G. W. 1989. 'Review: The State of the Art: Current Work in the Technology of Ancient Rome', *Classical Journal* 85: 63–80.

Hultsch, F. 1909. 'Eudoxos', in G. Wissowa (ed.), *Paulys Realencyclopädie der klassischen Altertumswissenschaft*, Stuttgart: Metzler, 6. 930–50.

Humphrey, J. W. 2006. *Ancient Technology*, Westport, Conn.: Greenwood Press.

Hunger, H. and D. Pingree 1989. *MUL.APIN: An Astronomical Compendium in Cuneiform*, Horn: Verlag Ferdinand Berger und Söhne.

Hunt, D. W. S. 1940–45. 'An Archaeological Survey of the Classical Antiquities of the Island of Chios Carried out between the Months of March and July, 1938', *Annual of the British School in Athens* 41: 29–52.

Hüttig, M. 2000. 'The Conical Sundial from Thyrrheion – Reconstruction and Error Analysis of a Displaced Antique Sundial', *Archive for History of Exact Sciences* 55: 163–76.

Jones, A. 2000. 'Calendrica I: New Callippic Dates', *Zeitschrift für Papyrologie und Epigraphik* 129: 141–58.

Jones, A. 2007. 'On Greek Stellar and Zodiacal Date-Reckoning', in J. M. Steele (ed.), *Calendars and Years: Astronomy and Time in the Ancient Near East*, Oxford: Oxbow Books, 149–67.

Jones, A. H. M. 1972. *The Criminal Courts of the Roman Republic and Principate*, Oxford: Blackwell.

Jones, Terry. 2008. *Barbarians*, part 3: 'The Brainy Barbarians', DVD, New York: Koch Vision.

Kannicht, R. 1986. *Tragicorum Graecorum fragmenta (TrGF)*, vol. 1, Göttingen: Vandenhoeck and Ruprecht.

Kannicht, R. 2004. *Tragicorum Graecorum fragmenta (TrGF)*, vol. 5, part 2, Göttingen: Vandenhoeck and Ruprecht.

Kidd, D. (ed.) 1997. *Aratus: Phaenomena*, Cambridge: Cambridge University Press.

Kidd, D. A. 1974. 'Hesperus and Catullus LXII', *Latomus* 33: 22–33.

Kienast, H. J. 1993. 'Untersuchungen am Turm der Winde', *Archäologischer Anzeiger* 1993: 271–75.

Kienast, H. J. 1997. 'The Tower of the Winds in Athens: Hellenistic or Roman?' in M. C. Hoff and S. I. Rotroff (eds), *The Romanization of Athens: Proceedings of an International Conference held at Lincoln, Nebraska, April 1996*, Oxford: Oxbow, 53–65.

Kienast, H. J. 2005. 'La Torre dei Venti di Atene', in E. Lo Sardo (ed.), *Eureka! Il Genio deglio Antichi*, Naples: Electa, 245–51.

Koch-Westenholz, U., 1995. *Mesopotamian astrology: An Introduction to Babylonian and Assyrian Celestial Divination*, Copenhagen: Carsten Niebuhr Institute of Near Eastern Studies, Museum Tusculanum Press, University of Copenhagen.

Künzl, E. 2005. *Himmelsgloben und Sternkarten: Astronomie und Astrologie in Vorzeit und Altertum*, Stuttgart: Theiss.

La Rocca, E. 1999. 'Pantheon (fase pre-adrianea)', in E. M. Steinby (ed.), *Lexicon Topographicum Urbis Romae*. Rome: Edizioni Quasar, 5, 280–83.

La Rocca, E., M. de Vos and A. de Vos 1994. *Pompei*, Milan: A. Mondadori.

Lane Fox, R. 1980. *The Search for Alexander*. London: Allen Lane.

Langholf, V. 1973. '*Hōra* = Stunde: zwei Beilage aus dem Anfang des 4. Jh. v. Chr.', *Hermes* 101: 382–84.

Lehmann, K. 1945. 'The Dome of Heaven', *Art Bulletin* 27: 1–27.

Lehoux, D. R. 2000. 'Parapegmata; or, Astrology, weather, and calendars in the ancient world', PhD thesis, University of Toronto.

Lehoux, D. R. 2004. 'Observation and Prediction in Ancient Astrology', *Studies in History and Philosophy of Science* 35: 227–46.

Lehoux, D. R. 2005. 'The Parapegma Fragments from Miletus', *Zeitschrift für Papyrologie und Epigraphik* 152: 125–40.

Lehoux, D. R. 2007. *Astronomy, Weather, and Calendars in the Ancient World: Parapegmata and Related Texts in Classical and Near-Eastern Societies*, Cambridge: Cambridge University Press.

Lewis, D. M. 1977. 'Never on the Day of the Agathos Daimon', *Classical Review* n.s. 27: 215–16.

Lewis, M. 2000. 'Theoretical Hydraulics, Automata, and Water Clocks', in Ö. Wikander (ed.), *Handbook of Ancient Water Technology*. Leiden: Brill, 343–69.

Lintott, A. 2004. 'Legal Procedure in Cicero's Time', in J. Powell and J. Paterson (eds), *Cicero the Advocate*. Oxford: Oxford University Press, 63–78.

Lippincott, K. 1999. *The Story of Time*, London: Merrell Holberton and National Maritime Museum.

Llewelyn, S. R. 1993. 'Did the Ptolemaic Postal Service Work to a Timetable?', *Zeitschrift für Papyrologie und Epigraphik* 99: 41–56.

Llewelyn, S. R. 1994. *New Documents Illustrating Early Christianity. Volume 7, A Review of the Greek Inscriptions and Papyri Published in 1982–83*. North Ryde, N.S.W.: Ancient History Documentary Research Centre, Macquarie University.

Lloyd, A. B. 1988. *Herodotus, book II*. Leiden: E. J. Brill.

Locher, K. 1989. 'A Further Hellenistic Conical Sundial from the Theatre of Dionysos in Athens', *Journal for the History of Astronomy* 20: 60–62.

L'Orange, H. P. 1947. *Apotheosis in Ancient Portraiture*, Oslo: H. Aschehoug.

Loraux, N. 2002. *The Divided City: On Memory and Forgetting in Ancient Athens*, trans. C. Pache and J. Fort, New York: Zone Books.

Magnuson, T. 2004. *The Urban Transformation of Medieval Rome, 312–1420*. Stockholm: Suecoromana VII.

Martin, R. 1965: *Manuel d'Architecture grecque, I: Matériaux et Techniques*, Paris: Éditions A. et J. Picard.

Mastrocinque, A. 2007. 'Late Antique Lamps with Defixiones', *Greek, Roman and Byzantine Studies* 47: 87–99.

McCluskey, S. 1977. 'The Astronomy of the Hopi Indians', *Journal for the History of Astronomy* 8: 174–95.

McCluskey, S. 1998. *Astronomies and Cultures in Early Medieval Europe*, Cambridge: Cambridge University Press.

McCluskey, S. 2000. 'The Inconstant Moon: Lunar Astronomies in Different Cultures', in *Archaeoastronomy* 15: 14–31.

McCluskey, S. 2004. 'Astronomy, Time, and Churches in the Early Middle Ages', in M.-T. Zenner (ed.), *Villard's Legacy: Studies in Medieval Technology, Science, and Art in Memory of Jean Gimpel*, Aldershot, Burlington: Ashgate, 197–210.

McClusky, S. 2007. 'Calendrical Cycles, the Eighth Day of the World, and the Orientation of English Churches', in C. Ruggles and G. Urton (eds), *Skywatching in the Ancient World: New Perspectives in Cultural Astronomy. Studies in Honour of Anthony F. Aveni*, Boulder, Co.: University Press of Colorado, 331–53.

Meiggs, R. and D. Lewis 1988. *A Selection of Greek Historical Inscriptions to the End of the Fifth Century B.C.*, Oxford: Clarendon Press.

Merry, W. W. 1879. *The Clouds*, Oxford: Clarendon Press.

Michels A. K. 1967. *The Calendar of the Roman Republic*, Princeton: Princeton University Press.

Mikalson, J. D. 1975. *The Sacred and Civil Calendar of the Athenian Year*, Princeton, N.J.: Princeton University Press.

Mills, A. A. 1993. 'Seasonal-Hour Sundials on Vertical and Horizontal Planes, with an Explanation of the Scratch Dial', *Annals of Science* 50: 83–93.

Mills, A. A. 2000a. 'Graeco–Roman Sundials, Part I: Dials Based on the Sphere', *Bulletin of the British Sundial Society* 12: 3–11.

Mills, A. A. 2000b. 'Graeco–Roman Sundials, Part II: Conical and Other Forms', *Bulletin of the British Sundial Society* 12: 64–70.

Mills, S. 1997. *Theseus, Tragedy, and the Athenian Empire*, Oxford: Clarendon Press.

Mitchell, S. 1976. 'Requisitioned Transport in the Roman Empire: A New Inscription from Pisidia', *Journal of Roman Studies* 66: 106–13.

Morgan, J. D. 1996. 'The Calendar and the Chronology of Athens', *American Journal of Archaeology* 100: 395.

Müller, J. W. 1991. 'Intercalary Months in the Athenian Dark-Age Period', *Schweizer Münzblätter* 41, 164: 85–89.

Müller, W. 2001. 'Andronikos (I)', in R. Vollkommer (ed.), *Künstlerlexikon der Antike*, vol. 1, Munich and Leipzig: K. G. Saur, 43–44.

Murray, G. 1933. *Aristophanes: A Study*, Oxford: Clarendon Press.

Neugebauer, O. 1941–43. 'Egyptian Planetary Texts', *Transactions of the American Philosophical Society* 32: 209–50.

Neugebauer, O. 1948. 'The Astronomical Origin of the Theory of Conic Sections', *Proceedings of the American Philosophical Society* 92: 136–38.

Neugebauer, O. 1975. *A History of Ancient Mathematical Astronomy*, Berlin, Heidelberg, New York: Springer-Verlag.

Neugebauer, O. and R. A. Parker 1969. *Egyptian Astronomical Texts*, vol. 3. Providence, R.I.: Brown University Press.

Nicholson, J. 1994. 'The Delivery and Confidentiality of Cicero's Letters', *Classical Journal* 90: 33–63.

Noble, J. V. and D. J. de Solla Price 1968. 'The Water Clock in the Tower of the Winds', *American Journal of Archaeology* 72: 345–55.

Oakley, S. P. (ed.) 2005. *A Commentary on Livy, Books VI–X: Volume 4, Book X*, Oxford: Oxford University Press.

Oleson, J. P. (ed.) 2008. *Handbook of Engineering and Technology in the Classical World*, Oxford: Oxford University Press.

Orlin, E. M. 2002. *Temples, Religion, and Politics in the Roman Republic*, Boston: Brill Academic.

Osborne, M. J. 2000. 'Philinos and the Athenian Archons of the 250s BC', in P. FlenstedJensen, T. H. Nielsen and L. Rubinstein (eds), *Polis and Politics: Studies in Ancient Greek History Presented to Mogens Herman Hansen on his Sixtieth Birthday, August 20, 2000*, Copenhagen: Museum Tusculanum Press, University of Copenhagen, 507–20.

Osborne, M. J. 2003. 'The Athenian Archon Diomedon and his Successors', *Zeitschrift für Papyrologie und Epigraphik* 143: 95–100.

Oudet J.-F. 1992. 'Le Panthéon de Rome à la lumière de l'equinoxe', in *Readings in Archaeoastronomy: Papers presented at the International Conference: Current Problems and Future of Archaeoastronomy, held at the State Archaeological Museum in Warsaw, 15–16 November 1990*, Warsaw: State Archaeological Museum, 25–52.

Parker, J. and D. Parker 1991. *Parkers' Astrology*, London: Dorling Kindersley.

Parker, R. A. 1974. 'Ancient Egyptian Astronomy', in F. R. Hodson (ed.), *The Place of Astronomy in the Ancient World*, London: The British Academy and Oxford University Press, 51–65.

Parker, R. A. and W. H. Dubberstein 1956. *Babylonian Chronology 626 B.C. – A.D. 75*, Providence, R.I.: Brown University Press.

Parsons, E. C. (ed.), 1936. *Hopi Journal of Alexander M. Stephen*, New York: Columbia University Press.

Pattenden, P. 1979. 'Sundials in Cetius Faventinus', *Classical Quarterly* 29: 203–12.

Pattenden, P. 1981. 'A Late Sundial at Aphrodisias', *Journal of Hellenic Studies* 101: 101–12.

Pattenden, P. 1987. 'When Did Guard Duty End? The Regulation of the Night Watch in Ancient Armies', *Rheinisches Museum* 130: 164–74.

Pedersen, S. and R. Hannah 2002. 'Celestial Dynamics at the Crossroads: Proclus' Reassessment of Plato in the Light of Empirical Science', *Antichthon* 36: 65–79.

Petrakos, B. Ch. 1968. *O Oropos kai to Ieron tou Amphiaraou*, Athens: En Athenias.

Picard, G. 1970. *Roman Painting*, London: Elek.

Poole, R. 1998. *Time's Alteration: Calendar Reform in Early Modern England*, London: UCL Press.

Potamianos, I. 2000. *To phōs stē Vyzantinē ekklēsia*, Thessalonike: University Studio Press.

Powell, J. E. 1940. 'Greek Timekeeping', *Classical Review* 54: 69–70.

Preisendanz, K. (ed.) 1973. *Papyri Graecae Magicae: die griechischen Zauberpapyri*, Stuttgart: B.G. Teubner.

Price, D. J. de Solla 1957. 'Precision Instruments: To 1500', in C. Singer, E. J. Holmyard, A. R. Hall and T. I. Williams (eds), *A History of Technology, Volume III:*

From the Renaissance to the Industrial Revolution c. 1500–c. 1750. Oxford: Clarendon Press, 582–619.

Price, D. J. de Solla 1969. 'Portable Sundials in Antiquity, Including an Account of a New Example from Aphrodisias', *Centaurus* 14: 242–66.

Price, D. J. de Solla 1974. 'Gears from the Greeks', *Transactions of the American Philosophical Society* 64, 7: 1–70.

Price, D. J. de Solla 1975. 'Clockwork before the Clock and Timekeepers before Timekeeping', in J. T. Fraser and N. Lawrence (eds), *The Study of Time II: Proceedings of the Second Conference of the International Society for the Study of Time, Lake Yamanaka, Japan*. New York: Springer-Verlag, 367–80.

Price, S. 1999. *Religions of the Ancient Greeks*, Cambridge: Cambridge University Press.

Pritchett, W. K. 1946. 'Months in Dorian Calendars', *American Journal of Archaeology* 50: 358–60.

Pritchett, W. K. 1947. 'Julian Dates and Greek Calendars', *Classical Philology* 42: 235–43.

Pritchett, W. K. 1999. 'Postscript: The Athenian Calendars', *Zeitschrift für Papyrologie und Epigraphik* 128: 79–93.

Pritchett, W. K. 2001. *Athenian Calendars and Ekklesias*, Amsterdam: J. C. Gieben.

Pritchett, W. K. and O. Neugebauer 1947. *The Calendars of Athens*, Cambridge, Mass.: Harvard University Press.

Pritchett W. K. and B. L. van der Waerden 1961. 'Thucydidean Time-Reckoning and Euctemon's Seasonal Calendar', *Bulletin de Correspondance Hellénique* 85: 17–52.

Ramsay, A. M. 1920. 'A Roman Postal Service under the Republic', *Journal of Roman Studies*, 10: 79–86.

Ramsay, A. M. 1925. 'The Speed of the Roman Imperial Post', *Journal of Roman Studies* 15: 60–74.

Ramsay, W. M. 1904. *The Letters to the Seven Churches of Asia, and their Place in the Plan of the Apocalypse*, 4th ed., London: Hodder and Stoughton.

Rayet, G. 1875. 'Cadrans solaires coniques', *Annales de chimie et de physique*, 52–86.

Rehm, A. 1913. *Griechische Kalender, iii. Das Parapegma des Euktemon*, Heidelberg: Carl Winter.

Rehm, A. 1941. *Parapegmastudien*, Munich: Verlag der Bayerischen Akademie der Wissenschaften.

Rehm, A. 1949. 'Parapegma', in A. Pauly, G. Wissowa and W. Kroll (eds), *Real-Encyclopedie des klassischen Altertumswissenschaft* 18: 1295–366. Stuttgart: Metzler.

Reiche, H. A. T. 1989. 'Fail-safe Stellar Dating: Forgotten Phases', *Transactions of the American Philological Association* 119: 37–53.

Remijsen, S. 2007. 'The Postal Service and the Hour as a Unit of Time in Antiquity', *Historia* 56: 127–40.

Richardson, W. F. 2004. *Numbering and Measuring in the Classical World*, Bristol: Bristol Phoenix Press.

Robertson, D. S. 1940. 'The Evidence for Greek Timekeeping', *Classical Review* 54: 180–82.

Rochberg, F. 2004. *The Heavenly Writing: Divination, Horoscopy, and Astronomy in Mesopotamian Culture*, Cambridge: Cambridge University Press.

Rogers, B. B. 1924. *Aristophanes*, vol. II, London: Heinemann.

Rose, H. J. 1970. 'Clocks', and 'Time-reckoning', in N. G. L. Hammond and

H. H. Scullard (eds), *The Oxford Classical Dictionary*, 2nd ed., Oxford: Clarendon Press, 253, 1075–76.

Roseman, C. H. (ed.) 1994. *Pytheas of Massalia, On the Ocean*, Chicago: Ares Publishers.

Rowland, I. D. and T. N. Howe (eds) 1999. *Vitruvius: Ten Books on Architecture*, Cambridge: Cambridge University Press.

Ruggles, C. L. N. 1999. *Astronomy in Prehistoric Britain and Ireland*, New Haven, Conn.: Yale University Press.

Salt, A. and E. Boutsikas 2005. 'Knowing When to Consult the Oracle at Delphi', *Antiquity* 79: 564–72.

Sambursky, S. 1956. *The Physical World of the Greeks*, London: Routledge and Paul.

Sambursky, S. and S. Pines 1971. *The Concept of Time in Late Neoplatonism*, Jerusalem: Israel Academy of Sciences and Humanities, Section of Humanities.

Samuel, A. E. 1972. *Greek and Roman Chronology*, Munich: C.H. Beck'sche Verlagsbuchhandlung.

Schaefer, B. E. 2005. 'The Epoch of the Constellations on the Farnese Atlas and the Origin in Hipparchus's Lost Catalogue', *Journal for the History of Astronomy* 36: 1–29.

Schaldach, K. 1998. *Römische Sonnenuhren: eine Einführung in die antike Gnomonik*, Frankfurt am Main: H. Deutsch.

Schaldach, K. 2004. 'The Arachne of the Amphiareion and the Origin of Gnomonics in Greece', *Journal for the History of Astronomy* 35: 435–45.

Schaldach, K. 2006. *Die antiken Sonnenuhren Griechenlands: Festland und Peloponnes*, Frankfurt am Main: Deutsch.

Schiaparelli, G. 1926. *Scritti sulla Storia della Astronomia Antica*, i/2. Bologna: Zanichelli.

Schilling, R. 1977. *Pline l'Ancien, Histoire naturelle, Livre VII*, Paris: Les Belles Lettres.

Seabrook, J. 2007. 'Fragmentary Knowledge: Was the Antikythera Mechanism the World's First Computer?', *The New Yorker* 14 May.

Seely, F. A. 1888. 'The Development of Time-Keeping in Greece and Rome', *American Anthropologist* 1: 25–50.

Sefrin-Weis, H. 2007. 'Review of Schaldach 2006', in *Aestimatio* 4: 29–39.

Sellers, J. B. 1992. *The Death of Gods in Ancient Egypt*, Harmondsworth: Penguin.

Shelton, J.-A. 1998. *As the Romans Did: A Sourcebook in Roman Social History*, 2nd ed., New York: Oxford University Press.

Smyth, A. P. 2002. *The Medieval Life of King Alfred the Great: A Translation and Commentary on the Text Attributed to Asser*, Basingstoke and New York: Palgrave.

Sommerstein, A. H. (ed.) 1982. *Aristophanes: Clouds*. Warminster: Aris & Phillips.

Sorabji, R. 1983. *Time, Creation and the Continuum: Theories in Antiquity and the Early Middle Ages*, London: Duckworth.

Soubiran, J. 1969. *Vitruve: de l'Architecture, Livre IX*. Paris: Les Belles Lettres.

Spalinger, A. 1991. 'Remarks on an Egyptian Feast Calendar of Foreign Origin', *Studien zur altägyptischen Kultur* 18: 349–73.

Sperling, G. 1999. *Das Pantheon in Rom: Abbild und Mass des Kosmos*, Neuried: Ars Una.

Stamper, J. W. 2005. *The Architecture of Roman Temples: The Republic to the Middle Empire*, Cambridge: Cambridge University Press.

Stern, S. 2000. 'The Babylonian Calendar at Elephantine,' *Zeitschrift für Papyrologie und Epigraphik* 130: 159–71.

Stern, S. 2008. 'The Babylonian Month and the New Moon: Sighting and Prediction', *Journal for the History of Astronomy* 39: 19–42.

Stevenson, W. H. (ed.) 1959. *Asser's Life of King Alfred, together with the Annals of Saint Neots, erroneously ascribed to Asser*, Oxford: Clarendon Press.

Stierlin, H. 1984. *Hadrien et l'Architecture romaine*, Paris: Payot.

Stierlin, H. 1986. *L'Astrologie et le pouvoir: de Platon à Newton*, Paris: Payot.

Stocks, J. L. 1930. *De Caelo*, in W. D. Ross (ed.), *The Works of Aristotle*, Volume II, Oxford: Clarendon Press.

Stott, C. 1991. *Celestial Charts: Antique Maps of the Heavens*, London: Studio Editions.

Stuart, J. and N. Revett 1762/2008. *The Antiquities of Athens: Measured and Delineated*, vol. I., new ed., London: Priestley and Weale; New York: Princeton Architectural Press.

Symons, S. 1998. 'Shadow Clocks and Sloping Sundials of the Egyptian New Kingdom and Late Period: Usage, Development and Structure', *Bulletin of the British Sundial Society* 98: 30–36.

Taub, L. 1997. 'The Rehabilitation of Wretched Subjects', *Early Science and Medicine* 2: 74–87.

Taub, L. 2001. 'Destini della scienza greca: eredità e longevità degli strumenti scientifici', in S. Settis (ed.), *I Greci: Storia, Cultura, Arte, Società*, 4 vols, Turin: Einaudi, 3: 889–930.

Taub, L. 2002. 'Instruments of Alexandrian Astronomy: The Uses of the Equinoctial Rings', in C. J. Tuplin and T. E. Rihll (eds), *Science and Mathematics in Ancient Greek Culture*, Oxford: Oxford University Press, 133–49.

Taub, L. 2003. *Ancient Meteorology*, London: Routledge.

Taub, L. 2006. 'Are Orreries "Newtonian"? A Consideration of the Material, Textual and Pictorial Evidence', in L. Taub and F. Willmoth (eds), *The Whipple Museum of the History of Science: Instruments and Interpretations, to Celebrate the 60th Anniversary of R. S. Whipple's Gift to the University of Cambridge*, Cambridge: Cambridge University Press, 403–25.

Taverner E. 1918. 'The Roman Farmer and the Moon', *Transactions of the American Philological Association* 49: 67–82.

Theodossiou, E. T. and E.-M. Kalyva. 2002. 'The Ancient Clepsydra of Athens', *Bulletin of the British Sundial Society* 14: 162–66.

Thomas, E. 1997. 'The Architectural History of the Pantheon in Rome from Agrippa to Septimius Severus via Hadrian', *Hephaistos* 15: 163–86.

Thomas, E. 2004. 'From the Pantheon of the Gods to the Pantheon of Rome', in R. Wrigley and M. Craske, *Pantheons: Transformations of a Monumental Idea*, Aldershot, UK and Burlington, Vt.: Ashgate, 11–34.

Thomas, R. 1989. *Oral Tradition and Written Record in Classical Athens*, Cambridge: Cambridge University Press.

Toomer, G. J. 1998. *Ptolemy's Almagest*, Princeton, N.J.: Princeton University Press.

Toomer, G. J. 2003. 'Astronomy', in S. Hornblower and A. Spawforth (eds), *The Oxford Classical Dictionary*, 3rd ed., revised. Oxford: Oxford University Press, 196–98.

Tracy, S. V. 1990. *Attic Letter-cutters of 229 to 86* B.C., Berkeley: University of California Press.

Trebilco, P. 2006. 'Christian Communities in Western Asia Minor into the Early

Second Century: Ignatius and Others as Witnesses against Bauer', *Journal of the Evangelical Theological Society* 49: 17–44.
Trevarthen, D. 2000. 'Illuminating the Monuments: Observation and Speculation on the Structure and Function of the Cairns at Balnuaran of Clava', *Cambridge Archaeological Journal* 10: 295–315.
Trümpy, C. 1997. *Untersuchungen zu den altgriechischen Monatsnamen und Monatsfolgen*, Heidelberg: C. Winter.
Turetzky, P. 1998. *Time*, London: Routledge.
Turner, A. J. 1989. 'Sun–dials: History and Classification', *History of Science* 27: 303–18.
Turner, A. J. (ed.) 1990. *Time*, The Hague: Tijd voor Tijd Foundation.
Turner, A. J. 1993. *Of Time and Measurement: Studies in the History of Horology and Fine Technology*, Aldershot: Variorum.
Turner, A. J. 1994. *Mathematical Instruments in Antiquity and the Middle Ages: An Introduction*, London: Vade-Mecum.
Valev, P. 2004. 'Mathematical and Astronomical Study of a Roman Sundial from Kabyle', *Archaeologica Bulgarica* 8: 55–62.
van Cleempoel, K. (ed.) 2005. *Astrolabes at Greenwich: A Catalogue of the Astrolabes in the National Maritime Museum, Greenwich*. Oxford: Oxford University Press.
van der Waerden, B. L. 1952–53. 'History of the Zodiac', *Archiv für Orientforschung* 16: 216–30.
van der Waerden, B. L. 1988. *Die Astronomie der Griechen: Eine Einführung*, Darmstadt: Wissenschaftliche Buchgesellschaft.
Veuve, S. 1982. 'Cadrans solaires gréco-bactriens à Aï Khanoum (Afghanistan)', *Bulletin de Correspondance Hellénique* 106: 23–51.
Virgili, P. 1999. 'Pantheon (età adrianea)', in E. M. Steinby (ed.), *Lexicon Topographicum Urbis Romae*, Rome: Edizioni Quasar, 5. 284–85.
Voisin J.-L. 1987. 'Exoriente Sole (Suétone, Ner. 6). D'Alexandrie à la Domus Aurea', in *L'Urbs: Espace urbain et Histoire (1er siècle av. J.-C.–IIIe siècle ap. J.-C.). Actes du colloque international organisé par le Centre national de la recherche scientifique et l'École française de Rome (Rome, 8–12 mai 1985)*, Rome: École française de Rome, 509–41.
von Boeselager, D. 1983. *Antike Mosaiken in Sizilien: Hellenismus und römische Kaiserzeit, 3. Jahrhundert v. Chr.–3. Jahrhundert n. Chr.*, Rome: Giorgio Bretschneider.
von Bomhard, A.-S. 1999. *The Egyptian Calendar: A Work for Eternity*, London: Periplus.
von Freeden, J. 1983. ΟΙΚΙΑ ΚΥΡΡΗΣΤΟΥ: *Studien zum sogenannten Turm der Winde in Athen*, Rome: Giorgio Bretschneider.
Walbank, M. B. 1982. 'Regulations for an Athenian Festival', *Studies in Attic Epigraphy, History and Topography, presented to Eugene Vanderpool. Hesperia Supplements*, vol. 19, 173–82.
Ward-Perkins, J. B. 1979. *Roman Architecture*, Electa: Milan.
West, J. A. 1991. *The Case for Astrology*, London: Arkana.
West M. L. 1978. *Hesiod: Works and Days*, Oxford: Clarendon Press.
West, M. L. 1987. *Euripides, Orestes*, Warminster: Aris & Phillips.
West, M. L. 1990. *Aeschyli Tragoediae cum incerti poetae Prometheo*, Stuttgart: B. G. Teubner.
West, S. 1973. 'Cultural Interchange over a Water–clock', *Classical Quarterly* 23: 61–64.

Westgate, R. 2000. '*Pavimenta atque emblemata vermiculata*: Regional Styles in Hellenistic Mosaic and the First Mosaics at Pompeii', *American Journal of Archaeology* 104: 255–75.

White, K. D. 1984. *Greek and Roman Technology*, Ithaca, N.Y.: Cornell University Press.

Whitehead, D. (ed.) 2002. *Aineias the Tactician: How to Survive Under Siege*, London: Bristol Classical Press.

Williams, J. Forthcoming. 'Puaka and Matariki', *He Pukenga Korero: A Journal of Maori Studies*.

Wilson Jones M. 2000. *Principles of Roman Architecture*, New Haven, Conn. and London: Yale University Press.

Wright, M. T. 2000. 'Greek and Roman Portable Sundials: An Ancient Essay in Approximation', *Archive for History of Exact Sciences* 55: 177–87.

Wright, M. T. 2002. 'A Planetarium Display for the Antikythera Mechanism', *The Horological Journal* 144: 169–73, 193.

Wright, M. T. 2003a. 'Epicyclic Gearing and the Antikythera Mechanism, Part 1', *Antiquarian Horology* 27: 270–79.

Wright, M. T. 2003b. 'In the Steps of the Master Mechanic', in *He archaia Hellada kai ho Synchronos Kosmos*, Patras: University of Patras, 86–97.

Wright, M. T. 2004. 'The Scholar, the Mechanic and the Antikythera Mechanism: Complementary Approaches to the Study of an Instrument', *Bulletin of the Scientific Instrument Society* 80: 4–11.

Wright, M. T. 2005a. 'The Antikythera Mechanism: A New Gearing Scheme', *Bulletin of the Scientific Instrument Society* 85: 2–7.

Wright, M. T. 2005b. 'Counting Months and Years: The Upper Back Dial of the Antikythera Mechanism', *Bulletin of the Scientific Instrument Society* 87: 8–13.

Wright, M. T. 2005c. 'Epicyclic Gearing and the Antikythera Mechanism, Part 2', *Antiquarian Horology* 29: 51–63.

Wright, M. T. 2005d. 'Il meccanismo di Anticitera: L'antica tradizione dei meccanismi ad ingranaggio', in E. Lo Sardo (ed.), *Eureka! Il Genio deglio Antichi*, Naples: Electa, 240–44.

Wright, M. T. 2006a. 'The Antikythera Mechanism and the Early History of the Moon-phase Display', *Antiquarian Horology* 29: 319–29.

Wright, M. T. 2006b. 'Understanding the Antikythera Mechanism', in *2nd International Conference on Ancient Greek Technology*, Athens: Technical Chamber of Greece, 49–60.

Young, S. 1939. 'An Athenian Clepsydra', *Hesperia* 8: 274–84.

Yunis, H. (ed.) 2001. *Demosthenes, On the Crown*, Cambridge: Cambridge University Press.

Zeeman, E. C. 1986. 'Gears from the Greeks', *Proceedings of the Royal Institution of Great Britain* 58: 137–56.

Ziolkowski, A. 1999. 'Pantheon', in E. M. Steinby (ed.), *Lexicon Topographicum Urbis Romae*, Rome: Edizioni Quasar, 4. 54–61.

INDEX

Note: *italic* page numbers denote references to figures/tables.